Environmental Science and Engineering

The ultimate goal of this series is to contribute to the protection of our environment, which calls for both profound research and the ongoing development of solutions and measurements by experts in the field. Accordingly, the series promotes not only a deeper understanding of environmental processes and the evaluation of management strategies, but also design and technology aimed at improving environmental quality. Books focusing on the former are published in the subseries Environmental Science, those focusing on the latter in the subseries Environmental Engineering.

Xueming Chen
Editor

Proceedings of the 2022 12th International Conference on Environment Science and Engineering (ICESE 2022)

Springer

Editor
Xueming Chen
College of Environment and Safety Engineering
Fuzhou University
Fuzhou, Fujian, China

ISSN 1863-5520 ISSN 1863-5539 (electronic)
Environmental Science and Engineering
ISBN 978-981-99-1380-0 ISBN 978-981-99-1381-7 (eBook)
https://doi.org/10.1007/978-981-99-1381-7

This Springer imprint is published by the registered company Springer Nature Singapore Pte Ltd.
The registered company address is: 152 Beach Road, #21-01/04 Gateway East, Singapore 189721, Singapore

Preface

The 2022 12th International Conference on Environment Science and Engineering (ICESE 2022) was successfully held online via Zoom during September 2–5, 2022. The conference has been successfully held in Bali, Bangkok, Beijing, Istanbul, Seoul, Barcelona, and Leuven.

ICESE 2022 is to bring together innovative academics and industrial experts in the field of Environment Science and Engineering to a common forum. The primary goal of the conference is to promote research and developmental activities in Environment Science and Engineering. Another goal is to promote scientific information interchange between researchers, developers, engineers, students, and practitioners working around the world. The conference will be held every year to make it an ideal platform for people to share views and experiences in Environment Science and Engineering and related areas.

The international conference was divided into four major parts: Online Opening Ceremony, Keynote Speeches, Oral Presentations, and Online Q&A Discussion. Details are as below: The whole event was conducted through Zoom. There were three conference speakers delivering their wonderful speeches and five sub-sessions where each author had 15 minutes for presentation and 5 minutes for questions.

The proceedings is a compilation of the accepted papers and represents an interesting outcome of the conference. All these papers were peer-reviewed by conference committee members and international experts, to guarantee their novelty, technical soundness, applicability, clarity of presentation, and relevance.

ICESE is an annual event which has been conducted since 2011. We will continue organizing this conference in the future to provide an effective platform for further exchange of new knowledge and potential collaboration in research areas of Environment Science and Engineering.

We are herewith extending our thankfulness to all the involved persons for actively contributing to the implementation of the conference and the technical program committee members who gave their valuable comments and suggestions for improving the papers. We also would like to thank the organizing committee, chairpersons, and sponsors for their valuable input in the organization of the conference.

Xueming Chen
Program Chair

Conference Committees

Conference Chairs

Jan Baeyens	Beijing University of Chemical Technology, China
Raf Dewil	KU Leuven, Belgium

Program Chairs

Franz Winter	Vienna University of Technology, Austria
Valentina Busini	Politecnico di Milano, Italy
Xueming Chen	Fuzhou University, China
Yifeng Xu	Wuhan University of Technology, China

Publicity Chair

Huili Zhang	Beijing University of Chemical Technology, China

International Steering Committee

Soteris Kalogirou	Cyprus University of Technology, Cyprus
Martin Schiemann	Bochum University, Germany
Rouzbeh Abbassi	Macquarie University, Sydney, Australia

Technical Committees

Abdeltif Amrane	University of Rennes 1, France
Snezhana Georgieva Gocheva-Ilieva	University of Plovdiv "Paisii Hilendarski", Bulgaria
Dimiter Velev	University of National and World Economy UNSS, Bulgaria
Kei Eguchi	Fukuoka Institute of Technology, Japan
Indah Prihartini	University of Muhammadiyah Malang, Indonesia

Norma Patricia Muñoz Sevilla	Instituto Politécnico Nacional, Mexico
Haliza Abdul Rahman	Universiti Putra Malaysia, Malaysia
Nilofar Asim	National University of Malaysia (UKM), Malaysia
Angel Torriero	Deakin University, Australia
Dj Borozan	Josip Juraj Strossmayer University of Osijek, Croatia
Germán Mazza	National University of Comahue, Argentina
Steve Dann Camargo Hinostroza	Universidad continental, Perú
Chang Choo Khean	Universiti Kuala Lumpur Malaysia France Institute, Malaysia
Chong Yong Ong	Director of Experiva Solutions, Shah Alam, Selangor, Malaysia
Franz Winter	Vienna University of Technology, Austria
Valentina Busini	Politecnico di Milano, Italy
Abdullah Al-fawwaz	Al al-Bayt University, Jordan
David Quanrud	University of Arizona, America
Sabri Berhail	University Center of Mila, Algeria
Mahmoud Nasr	Egypt-Japan University of Science and Technology (EJUST), Alexandria, Egypt
Xueming Chen	Fuzhou University, China
Jing Zhao	Guangdong University of Science and Technology, China
Siyu Zhang	Tsinghua University, China

Contents

Wastewater Treatment and Soil Remediation

Evaluation of the Efficiency of Silica Removal in Effluents with *(Zea Mays Var. Saccharata)* in the Santa Rosa 94–1 Mining Concession in C.C Llocllapampa, Jauja-Peru 3
Deysi Aliaga Zevallos, Claudio Lopez Tantavilca, Paola Jeremias Espinoza, and Nelida Tantavilca Martinez

Initiation and Recovery of Nitrosation in Zeolite Bioreactors 17
Jiajun Luo, HaiYan Fu, YiCheng Wu, YuanMao Zheng, and ShuGuang Wang

Immobilized Anaerobic Digestion of Molasses-Based Distillery Wastewater in Moving Bed Biofilm Reactor 29
Michelle Almendrala, Shekinah Mae Villaflor, Zhane Ann Tizon, Bonifacio Doma, and Ralph Carlo Evidente

Effects of Straw Returning on Paddy Soil Environmental Remediation 37
Chunying Li and Chunlong Zhang

Expression and Characterization of Dichloromethane Dehalogenase for Dichloromethane Degradation 50
Yicheng Wu, Chen Yu, Haiyan Fu, and Huichao Yao

Resource Management and Ecological Environment Protection

Optimal Solution for Increasing Groundwater Pumping by Integrating MODFLOW-USG and Particle Swarm Optimization Algorithm: A Case Study of Wadi El-Natrun, Egypt 59
Ahmed M. Saqr, Mahmoud Nasr, Manabu Fujii, Chihiro Yoshimura, and Mona G. Ibrahim

Estimating VKT on Rural Roads for Carbon Emission Estimation 74
Peng Zhang, Lifeng Xing, Lukai Zhang, and Zhenyu Li

Toxic Effects of Chromium (VI) on *Chlorella vulgaris* in Water 87
Zhouyi Zhou, Yicheng Wu, Binghui Liu, Chen Yu, and Haiyan Fu

Analysis of the Deforestation Process of Amazon from During 2001 to 2020 in Peru ... 94
Anali Chulluncuy Samaniego, Magaly Godys Avellaneda Ponce, and Steve Dann Camargo Hinostroza

Evaluation of a Reliable Method for Flash Flood Hazard Mapping in Arid Regions: A Case Study of the Gulf of Suez, Egypt ... 103
Mahmoud M. Mansour, Mahmoud Nasr, Manabu Fujii, Chihiro Yoshimura, and Mona G. Ibrahim

Numerical Study on Response Characteristics of Layered Rocks Under Tunnel Blasting ... 118
Yingdong Pan, Shuyong Li, Jian Guo, Feili Wu, and Xiaobin Ding

Clean Energy Technology and Building Energy Management

Prediction and Regression Analysis of Trans-Himalaya New Energy Development Trends Based on Logistic Model ... 131
Panpan Deng

Algae Biomass for Biofuel Production; Yield Analysis of *Chlorella Vulgaris* and *Scenedesmus Meyen* in Different Culture Media ... 144
Mona Alariqi, Wei Long, and Baba Imoro Musah

Hardware in the Loop and Experiment Validation for a PV Inverter Under Flexible Power Control ... 151
Achara Pichetjamroen, Nithiphat Teerakawanich, Phacharawat Chindamanee, and Natthawuth Somakettarin

Bioclimatic Architecture as a Design Basis for the Use of Renewable Energies and Sustainable Development, the Case of Sustainable Social Housing in Patagonia, Aysén, Chile ... 158
Samuel Carmona, Stefania Pareti, Loreto Rudolph, and David Flores

Simulation of the Use of Air-Source Heat Pumps in Different European Cities ... 164
Agata Ołtarzewska and Dorota Anna Krawczyk

Simulation Analysis of Dynamic Characteristics of Urban Central Cooling System ... 171
Dong-liang Li, Li Gong, Guo-qing Zhu, Guo-qiang Zhan, and Xu e-jie Wang

Author Index ... 181

Wastewater Treatment and Soil Remediation

Evaluation of the Efficiency of Silica Removal in Effluents with *(Zea Mays Var. Saccharata)* in the Santa Rosa 94–1 Mining Concession in C.C Llocllapampa, Jauja-Peru

Deysi Aliaga Zevallos[1](✉), Claudio Lopez Tantavilca[2], Paola Jeremias Espinoza[3], and Nelida Tantavilca Martinez[1]

[1] Facultad de Ingeniería de Minas, Universidad Continental, Huancayo, Peru
{71999480,ntantavilca}@continental.edu.pe
[2] Faculty of Industrial Engineering, Universidad Continental, Huancayo, Peru
70288199@continental.edu.pe
[3] Facultad de Ingeniería Ambiental, Universidad Continental, Huancayo, Peru
75492007@continental.edu.pe

Abstract. In recent years in Peru, there are few non-metallic mining companies, one of the main ones being the Santa Rosa 94–1 mining concession, which is located in the community of Llocllapampa Jauja - Peru, where the main resource is silica sand that is used in this place, although it also generates a negative impact on the siliceous sand washing operation process since it modifies the conditions of the surface water since they do not have a treatment system. The objective of the research is to determine the efficiency and the amount of organic coagulant (Zea mays var. Saccharata) for the removal of silica from the effluent in the mining concession. The characterization of the silica was carried out through petrographic analysis to know its composition, the effluent of the Total Suspended Solids (TSS) parameter was characterized, then it was designed factorially for 8 treatments with variation of the organic amendment, volume and time, which was applied to removal processes. After the analysis, they gave as a result that the non-metallic mining effluent generates a negative environmental impact on the ecosystem, comprising 15010 mg/L of TSS which exceeds the Maximum Permissible Limits, the combinations of organic amendment, effluent and time were developed, where in treatment 8 an average removal of 5,999 g/ml was obtained, which represents 88%, being the most efficient for silica removal.

Keywords: Removal · Silica · Effluent · Organic reagent

1 Introduction

SiO2, known as quartz, is the main constituent of silica sand, rocks that can be intrusive, extrusive and sedimentary; this sand has several industrial applications. It is given because of its chemical and physical stability, the only detrital mineral that is chemically pure, so it can be 100% SiO2, consisting of 46.7% of Silica and 53.3% of Oxygen [1],

X. Chen (Ed.): ICESE 2022, ESE, pp. 3–16, 2023.

for the extraction requires investments in large quantities, necessarily for the machinery and equipment that will be needed to move the sand, as well as to achieve degrees of purity and quality. Silica can be found in three main forms: pure, unconsolidated areas and consolidated rocks [1]. Silica is a non-metallic mineral that is extracted from the rocks where it is found in open-pit mining, being processed to the point of crushing to obtain a higher concentration of silica.

Globally, China is the largest producer of silica in 2021 with 6 million metric tons, followed by Russia [2]. Which, until today silica has been very important for each country, specifically in Peru, non-metallic mining has been growing gradually, being the department of Junin in the month of June in the year 2021 and 2022 there is a percentage of variance 202.2% [3]. Peru is a country of ancient mining tradition that has a geological potential, the presence of the Andes Mountains throughout the territory, constitutes our main source of mineral resources (MINEM) [4], besides having a great potential in non-metallic minerals also known as Industrial Minerals, such as silica, marble, diatomite (first producer in South America) and borates. Peru is considered one of the few countries in the world where large deposits of these minerals can be found.

The activities in the town of Llocllapampa are extremely important because it has caused problems to the natural environment due to the irrational exploitation of the quarry, which is affected day by day, increasing the problems and generating a great impact not only socially but also environmentally; specifically, in the operation of washing the silica sand, it modifies the surface water conditions because it flows through the Poncijalo stream, discharging its waters into the Mantaro River. On the other hand, it also affects the living beings that consume it, because the effluents of the Santa Rosa 94–1 Mining Company do not have a treatment area where the substances it contains can be eliminated to continue the cycle, so alternative solutions were established to reduce the high concentration of silica in the effluents, starting with project management, planning, execution, follow-up, control and closure, also the development of the engineering of the project with its respective stages, within which is the recognition of the area, sampling, laboratory analysis, implementation of technology based on organic reagents at various scales, performance monitoring and data processing, using factorial experimental designs, obtaining a true result of each process in execution, which benefits the community of Llocllapampa. Therefore, the objective is to determine the efficiency and the amount of organic reagent for the removal of silica from the effluent in the Santa Rosa 94–1 mining concession.

2 Materials and Methods

2.1 Materials

The following materials were used to determine the efficiency of organic coagulant removal (see Table 1).

A. **Organic coagulant (corn starch)**
   It is an organic compound of biological origin, the main ones being starch, cellulose, vegetable extracts, alginates, among others. Being an alternative with great potential because they are biodegradable and do not cause any harm to the environment [5].

Table 1. Materials used in the laboratory

Material	Quantity
Corn starch *(Zea mays var. Saccharata)*	2400 g
Effluent (wastewater with silica)	1L

B. **Effluent from Santa Rosa mining concession 94–1**
The effluents from the Santa Rosa 94–1 mining concession are the mixture of water with non-metallic mineral (silica) that is usually generated in a mining production process, which in many cases affects the environment in which it is developed.

2.2 Methodology

Location and Site Description. The "Santa Rosa 94–1" mining concession is owned by the farming community of Llocllapampa. The Santa Rosa 94–1 quarry is located in the community of Llocllapampa at 3496 masl, located in the central Andes of Peru, 52 km from the city of Huancayo. According to the mining rights summary of the Instituto Geológico, Minero y Metalúrgico (INGEMMENT), the concession covers an area of 300 hectares. In the Geological and Mining Cadastral Information System (GEOCATMIN) the Status of Right code is 020003094 mentioned in its Concession File [6] (see Fig. 1).

Fig. 1. Concesión Minera Santa Rosa 94–1. Source: [6]

The study area is located by geographic coordinates [7] (see Table 2). Politically, it is located in Ajocucho, Llocllapampa district, province of Jauja in the department of Junín (see Fig. 2). The Llocllapampa community has a diverse climate; from October to

Table 2. UTM coordinates PSAD

Vertex	North	East
1	8,692,000.00	434,000.00
2	8,691,000.00	434,000.00
3	8,691,000.00	432,000.00
4	8,693,000.00	432,000.00
5	8,693,000.00	433,000.00
6	8,692,000.00	433,000.00

Source: INGEMMET, 2022 [7]

May there is rainfall (8° - 20°) with precipitation and from May to September there is an absence of rain. The predominant crops are tubers, forage oats, and ichu in the higher elevations, and the surrounding population raises cattle [8].

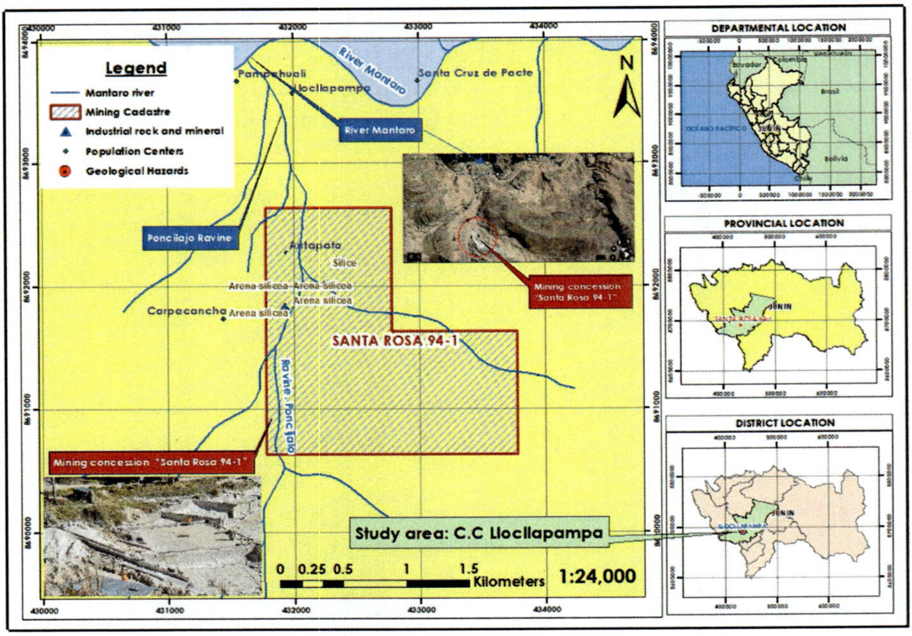

Fig.2. Location of the study area: "Santa 94–1" Mining Concession.

The Santa Rosa 94–1 mine consists of a process in which 50 people work in rotating shifts, the first process being drilling and blasting, which has a 10-min water dragging, which reaches the classification plant by granulometry of $1^{1/2}$ mm, 1 mm, 1/4 mm, the latter is used for construction, then it goes through double washing of silica 1/8 mm, and finally goes through a 1/16 mm washing process, to pass through drying ponds, where the effluent process does not have a treatment plant which comes to form a stream, which comes to flow into the Mantaro River directly contaminated (see Fig. 3).

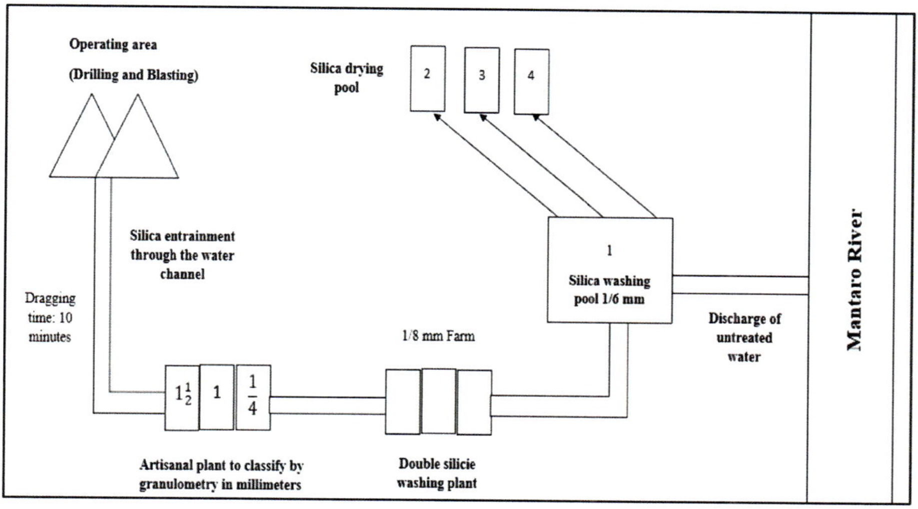

Fig. 3. Silica processing in the mining concession

Silica Characterization. Silicon dioxide, commonly known as silica, is a silicon oxide with formula SiO2 that is commonly found in nature as quartz. Due to its chemical and physical stability, it is the most abundant detrital mineral [1]. The silica extraction process in the Santa Rosa 94-1 mining concession is carried out with an open pit in the form of quarries, with benches of heights greater than 40 m and a mining platform (Fig. 4a). During the whole process the hydraulic force of water is used, which is supplied by channels coming from the "Poncijalo" stream, so that the granulated material deposited by gravity on the lower loading platform is driven through the channels to wash the broken material (Fig. 4b) [8], later the effluent, not having a treatment plant, comes to form a stream until it flows into the Mantaro river (Fig. 4c).

Fig. 4. During the visit to the Santa Rosa 94 - 1 mining concession in 2022, the silica deposit and the open pit activity were observed (a), also the path of the effluent was observed before entering the drying lagoon 1 for silica concentration (b) and finally the path of the effluent was observed (c).

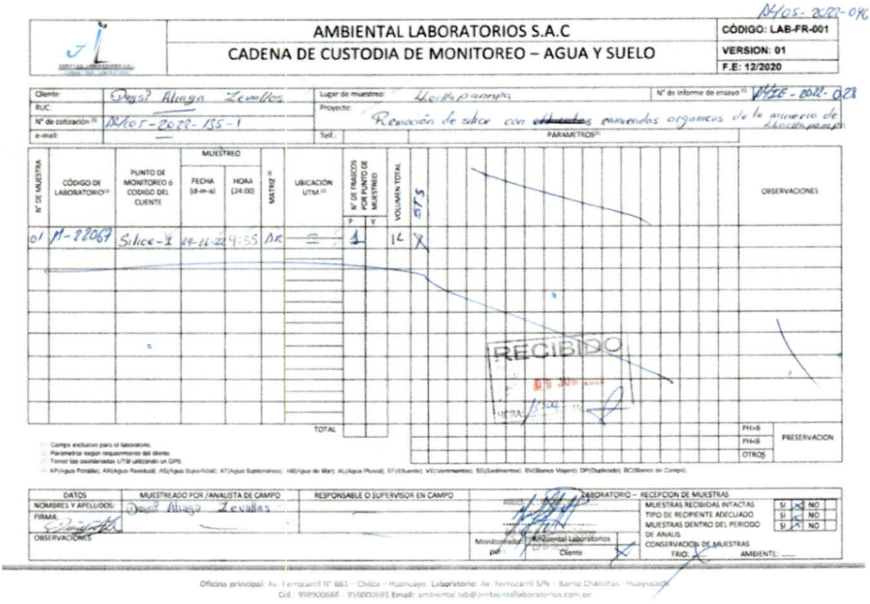

AMBIENTAL LABORATORIOS S.A.C
CADENA DE CUSTODIA DE MONITOREO – AGUA Y SUELO
CÓDIGO: LAB-FR-001
VERSION: 01
F.E: 12/2020

Cliente: Deysi Aliaga Zevallos
RUC:
N° de cotización: AL-COT-2022-135-1
e-mail:
Lugar de muestreo:
Proyecto: Remoción de sílice con enmiendas orgánicas de la minería de
Telf.:
N° de informe de ensayo:
PARAMETROS

N° DE MUESTRA	CÓDIGO DE LABORATORIO	PUNTO DE MONITOREO ó CODIGO DEL CLIENTE	MUESTREO FECHA (d-m-a)	MUESTREO HORA (24:00)	MATRIZ	UBICACIÓN UTM	N° DE FRASCOS POR PUNTO DE MUESTREO P	V	VOLUMEN TOTAL	STS	OBSERVACIONES
01	M-22067	Silice-1	19-11-22	9:55	AR		1		1L	X	

TOTAL
PRESERVACION: PH>8, PH<8, OTROS
RECIBIDO

DATOS	MUESTREADO POR /ANALISTA DE CAMPO	RESPONSABLE O SUPERVISOR EN CAMPO
NOMBRES Y APELLIDOS:	Deysi Aliaga Zevallos	
FIRMA:		
OBSERVACIONES		

LABORATORIO – RECEPCION DE MUESTRAS
MUESTRAS RECIBIDAS INTACTAS SI X NO
TIPO DE RECIPIENTE ADECUADO SI X NO
MUESTRAS DENTRO DEL PERIODO DE ANALIS SI X NO
CONSERVACION DE MUESTRAS FRIO: X AMBIENTE:
Monitoreado por: Ambiental Laboratorios / Cliente

Fig. 5. Chain of custody for effluent monitoring. Source: Ambiental Laboratorios S.A.C

Silica Characterization. To perform the characterization was carried out with the receipt of materials or equipment by the laboratory Environmental Laboratories S.A.C which provided us with the Chain of Custody of Water, where after it was collected a single sample "Silica 1", with a volume of 1L of effluent that was taken at 11°49′59. 0 "S 75°37′27.2" W, being a simple sampling at 9:55 a.m. being generated by the Santa Rosa 94–1 Mining Concession (see Fig. 5).

Silica Characterization

Wastewater Analysis. It was given through a SINGLE OR SPOT SAMPLE, according to the ANA (2016), this type of sample is also called discrete [9]. Which consisted of taking a portion of water in the area of 11°49′59.0 "S 75°37′27.2" W of the Santa Rosa 94–1 mining, with the volume of 1L of wastewater, which was analyzed the parameter of suspended solids in the laboratory "Ambiental Laboratorios S.A.C", with the SMEWW-APHA-AWWA-WEF Part 2540 D, 23 rd Ed. 2017 method whose description is Solids. Total Suspended Solids Dried at 103–105 °C (see Fig. 6).

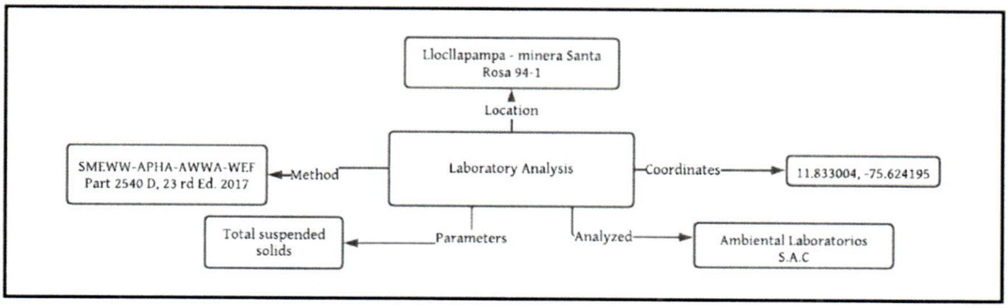

Fig. 6. Wastewater analysis process

Petrographic Analysis. Petrographic analysis helps us to describe and classify the composition of a sample with the main objective of analyzing the nature and origin of the sample by determining the type of rock, as well as the minerals and the degree of weathering [10].

a) Lithology

The Santa Rosa 94-1 mine is characterized by presenting a silica deposit, which is represented with a sequence of sedimentary and volcanic rocks, which is between the Neogene - Quaternary (NQ) period series. The white to gray silica is interbedded with the opaque yellow to dark orange sandstones due to the different sedimentary faces on which they were deposited (see Table 3).

Table 3. Prospecting for Industrial Rock and Mineral Resources in the Junín Region

Chronostratigraphic Column - Industrial Rocks and Minerals Resources In The Junín Region				
Eratema	**System**	**Series**	**Symbology**	**Alto Condoroma - Caylloma, Western Cordillera and Eastern Cordillera Lithostratigraphic Unit**
Cenozoic	CUATERNARIO	HOLOCENE	Q	Residual Deposit
				Biogenic Deposit
				Colluvial Deposit
				Alluvial Deposit
				Fluvial Deposit
				Chemical deposit, travertine
				Moraine deposits
				Glacial deposit
		PLEISTOCENE		Fluviglacial deposit
				Alluvial deposit
			NQ	
	NEOGEN	PLIOCENE		La Merced Formation
				Jauja Group-Mataula Formation
				Pacococha Formation
				Ushno Formation
		MYOCENE		Ipururo Formation
				Incahuasi Formation
				Millotingo Formation
				Yanacancha Formation
				Caudalosa Formation

Source: Chronostratigraphic column (Ingemmet, 2018) [11]

Factorial Design

For the analysis of silica removal will be performed with 8 experimental treatments.

$$2^3 = 8$$

$$A: \text{ Amount of coagulant/flocculant (reagent)} \quad \begin{matrix} t1: 30g \\ t2: 60g \end{matrix}$$

$$B: \text{ Sample volum} \quad \begin{matrix} v1: 200\ mL \\ v2: 400\ mL \end{matrix}$$

$$C: \text{ Sedimentation time} \quad \begin{matrix} t1: 1h \\ t2: 2h \end{matrix}$$

Table 4 shows the operationalization variables, where it is divided into 3 dimensions, being the quantity (c) to be used of 30 g and 60 g, in addition to the effluent volume (v) of 200 and 400 ml, and the sedimentation time (t) in a range of 1 and 2 h, which will be used for the combination of treatments.

Table 4. Variables and operationalization

Variables	Dimension	Indicators	Items
Organic Reagent	Quantity (g)	Amount of reagent to decrease the silica concentration in a certain sample volume to XX%	Amount of organic reagent dosed: c1: 30 g c2:60 g
Sample	Sample volume		Sample volume v1: 200 ml v2:400 ml
Sedimentation	Sedimentation time		Sedimentation time t1:1h t2:2h

Table 5 shows the experimental execution which consists of 8 treatments, which is made up of the amount of coagulant (c), volume of effluent used (v) and the time taken (t).

Table 5. Combinations of treatments

Experimental run	Combination of treatments
1	c1v1t1
2	c2v1t1

(continued)

Table 5. (*continued*)

Experimental run	Combination of treatments
3	c1v2t1
4	c2v2t1
5	c1v1t2
6	c2v1t2
7	c1v2t2
8	c2v2t2

3 Results

3.1 Parameters Evaluated with Respect to Water Quality

The parameter to evaluate the effluent is total suspended solids.

Total Suspended Solids. The results obtained by performing the analysis through the laboratory "Ambiental Laboratorios S.A.C", with the suspended solids parameter with the SMEWW-APHA-AWWA-WEF Part 2540 D, 23 rd Ed. 2017 method whose de-scription is Solids. Total Suspended Solids Dried at 103-105°C.

As established in the Supreme Decree 010–2010 of the Ministry of Environment [12], the results of the effluent in the **laboratory** exceed the maximum allowable limits of the effluent evaluated in the laboratory (see Table 6 and Table 7).

The Maximum Permissible Limit for the effluent from the Santa Rosa 94–1 mining concession is the concentration of substances in the liquid effluent from the activities carried out, which when exceeded causes damage to the health of the people living near the site and, above all, affects the ecosystem.

Table 6. Maximum permissible limits for the discharge of liquid effluents from mining and metallurgical activities.

PARAMETER	UNIT	Anytime Limit	Limit for Annual Average
pH		6–9	6–9
Total Suspended Solids	mg/L	50	25

Source:[12]

Table 7. Effluent Laboratory Results

Código	Ensayo	Resultado	Unidad
Sílice 1	Sólidos Totales en Suspensión	15010	mg/L

3.2 Laboratory Analysis Results

Amount of Silica. The results are presented from the laboratory analysis. This table shows the initial silica concentration without addition of organic reagents (See Table 8).

Silica Removal Analysis. This table shows the combinations performed and the average silica removal for each of them (See Table 9).

Table 8. Amount of initial silica

Initial Silica Quantity (0.017 g/mL)	
Sample volume	Silica Mass
200 mL	3.4085 g
400 mL	6.817 g

Table 9. Repeat treatments

Combination	Repetitions				Average g/ml
	1	2	3	4	
c1v1t1	1.701	1.750	1.634	1.715	1.700
c2v1t1	2.322	2.325	2.306	2.328	2.320
c1v2t1	3.409	3.405	3.400	3.418	3.408
c2v2t1	4.635	4.634	4.624	4.645	4.634
c1v1t2	2.405	2.405	2.410	2.404	2.406
c2v1t2	2.999	2.985	3.100	2.910	2.999
c1v2t2	4.781	4.785	4.785	4.770	4.780
c2v2t2	5.999	5.950	5.840	6.208	5.999

For the experimentation, disinfected and regulated laboratory equipment was used (see Fig. 7.a) in addition 8 different repetitions were performed (see Fig. 7b) as a result we obtained samples with removed silica (see Fig. 7c).

This table shows the percentage removal for each treatment combination. The higher the corn starch additive over a longer period of time, the higher the percentage removal of silica, with 88% removal efficiency in the treatments of 60 g in 200 ml with a time of 2 h and 60 g in 400 ml with a time of 2 h (see Table 10).

Fig. 7. The experimental process was observed in the laboratory, where materials were used for the testing process (a), the 8 treatments were prepared (b) and finally the silica removal results were observed (c).

Table 10. This table shows the combinations carried out and the average silica removal for each of them

Experimental run	Combination of treatments	Amount of silica	Amount of removal g/ml	% Removal
1	c1v1t1/30g-200 ml-1h	3.4085 g	1.700	50%
2	c2v1t1/60g-200 ml-1h	3.4085 g	2.320	68%
3	c1v2t1/30g-400 ml-1h	6.817 g	3.408	50%
4	c2v2t1/60g-400 ml-1h	6.817 g	4.634	68%
5	c1v1t2/ 30g-200 ml-2h	3.4085g	2.406	70%
6	c2v1t2/60 g-200 ml-2h	3.4085 g	2.999	88%

(*continued*)

Table 10. (*continued*)

Experimental run	Combination of treatments	Amount of silica	Amount of removal g/ml	% Removal
7	c1v2t2/30g-400 ml-2h	6.817 g	4.780	70%
8	c2v2t2/60 g-400 ml-2h	6.817 g	5.999	88%

4 Discussion

4.1 Organic Amendment and Fuculant -Flocculant Obtained from Nopal Cactus

The study was carried out using organic coagulant (corn starch) where eight trials of 30 g and 60 g of coagulant were carried out, with a time of 1 h and 2 h in volumes of 100 ml and 200 ml, obtaining a removal efficiency of 50%, 68%, 70% and 88%, this experiment was conducted in the laboratory of the Continental University, where the test system was used in beakers to simulate the coagulation process, being where the maximum removal of 5. 999 g/ml with test eight, where 60 gr of coagulant was added in 400 ml with a time of 2h having a percentage of 88% in the total removal, being the most optimal treatment for the removal of silica to the test carried out of suspended solids in Minera Santa Rosa 94–1, Thus, the use of "Opuntia ficus-indica" mucilage should be implemented for a better efficiency, since it presents a great effectiveness of 80% as treatment in the effluent of the mine belonging to the concessions of Compañía Minera Colquirrumi S. A.[13].

5 Conclusion

- The wastewater effluent from the Santa Rosa 94–1 non-metallic mining company was analyzed because it does not have a treatment system for its effluents discharged into the receiving body (Mantaro River), as evidenced in the laboratory results, exceeding the Maximum Permissible Limits for the discharge of liquid effluents from mining-metallurgical activities, which generates a negative environmental impact and affects the ecosystem.
- After performing the 8 treatments with different combinations of organic amendment, amount of effluent and time, it was observed that after the repetitions of each treatment, in this case 4, the most optimal was treatment 8 (c2v2t2), where 60 g of amendment were used in a volume of 400 ml with a time of 2 h, obtaining as a result an average removal of 5.999 g/ml.
- It was observed that after the tests carried out for treatments 6 (60 g of amendment in 200 ml with a time of 2 h) and 8 (60 g of amendment in 400 ml with a time of 2 h), the results obtained for silica removal were 88%, being the most optimal in the Santa Rosa 94–1 mining concession.
- The application of the natural flocculant in the samples showed that corn starch is efficient in the removal of silica in the effluent of the non-metallic mining company.

References

1. General directorate of mining development. SILICA MARKET PROFILE. Economy secretary, pp. 3–5 (2021)
2. Statista Homepage. https://es.statista.com/estadisticas/600220/paises-lideres-en-la-produccion-de-silicio-a-nivel-mundial/#statisticContainer. Accessed 17 Jun 2022
3. Ministry of Energy and Mines. MINING STATE BULLETIN, 2nd edn. Publisher: Peru (2022)
4. Ministry of energy and mines. https://www.minem.gob.pe/_detalle.php?idSector=1&idTitular=159&idMenu=sub149&idCateg=159#:~:text=Destacados%20de%20la%20Industria%20Minera,de%20dep%C3%B3sitos%20minerales%20del%20mundo. Accessed 17 Jun 2022
5. Delmester, C.: Coagulation with starchy corn starch for the removal of iron, turbidity and color in the confined aquifer of Yantaló. Peru (2019)
6. Geological, Mining and Metallurgical Institute(INGEMMET) Homepage https://portal.ingemmet.gob.pe/web/guest/sidemcat?CodDM=020003094&TipoDoc=0. Accessed 10 Jun 2022
7. Geological, Mining and Metallurgical Institute(INGEMMET) Homepage, "https://portal.ingemmet.gob.pe/igm-sidemcat-portlet/ver-resumen?pCodigo_DM=020003094", last accessed 2022/06/10
8. Omar, B., Ivan, H.: Optimization and modernization in the process of obtaining increased production in quarry Santa Rosa 94-ICC. Peru (2014)
9. National water authority. National Protocol for Monitoring the Quality of Surface Water Resources. Peru (2016)
10. Micromatco. Petrographic analysis of aggregates standard ASTM C-295. Colombia (2018)
11. Mario,C., May, L.: Prospecting of rock resources and industrial minerals in the junín region. Lima-Perú (2018)
12. Supreme Decree No. 010–2010-Minam. Approve maximum permissible limits for the discharge of liquid effluents from mining - metallurgical activities 21 Aug 2010
13. Jhan Pierre, C., Jhonatan, C.: Evaluation of the mucilage of opuntia ficus-indica in the reduction of metals in acid drainage of the quebrada honda mine-colcquirrumi S.A. Peru (2018)

Initiation and Recovery of Nitrosation in Zeolite Bioreactors

Jiajun Luo[1], HaiYan Fu[1], YiCheng Wu[1], YuanMao Zheng[1], and ShuGuang Wang[2](✉)

[1] Key Laboratory of Environmental Biotechnology (XMUT), Fujian Province University, Xiamen University of Technology, Xiamen 361024, China
[2] Shandong University, Qingdao 266237, China
wsg@sdu.edu.cn

Abstract. Stable partial nitrosation is important for processes based on anaerobic ammonia oxidation. In this study, an up-flow sludge fixed-bed reactor filled with natural zeolites at the bottom was used, under the condition that no organic carbon source is added, after gradually increasing the influent ammonia nitrogen load (ALR). At d 21, stable partial nitrosation was successfully initiated. The results showed that the influent ammonia nitrogen load was 0.5451 kg/(m^3 d) at an HRT of 16 h. The nitrite accumulation rate (NAR) of the system was 91.52% at this time, and the nitrite yield was 0.2384 kg/(m^3 d), and the effluent met the anaerobic ammonia oxidation influent requirements. After reducing the influent ammonia nitrogen concentration to 250 mg/L, the system nitrosation stability was disrupted and the NAR dropped to a minimum of 21.35%, but after restoring the influent ammonia nitrogen concentration, the system NAR was restored to 92.79% after only 6 d. The inhibition of nitrite oxidising bacteria (NOB) by ammonia nitrogen loading is an important reason why stable nitrification is achieved. Zeolite dosing allows the system to quickly recover nitrosation performance when the influent ammonia nitrogen concentration fluctuates. Stable partial nitrosation anaerobic ammonia oxidation process laid the foundation for stable application in engineering.

Keywords: Nitrosation · Ammonia nitrogen load · Anaerobic ammonia oxidation · Zeolite

1 Introduction

The traditional biological denitrification process is carried out by microorganisms such as ammoniacal bacteria, nitrifying bacteria and denitrifying bacteria, while denitrifying bacteria are heterotrophic bacteria and denitrification requires organic carbon in the wastewater as a carbon source as well as providing an electron donor [1]. With the improvement of people's quality of life and the development trend of carbon neutrality, the total amount of wastewater is increasing, while the C/N ratio in wastewater is decreasing, and the carbon source of denitrifying bacteria is insufficient, resulting in additional carbon sources for denitrification treatment in wastewater treatment plants,

X. Chen (Ed.): ICESE 2022, ESE, pp. 17–28, 2023.

which greatly increases the operating costs of wastewater treatment plants and generates some additional carbon emissions [1].

Compared with traditional nitrification and nitrosation, anaerobic ammonia oxidation is a nitrogen removal process that is consistent with carbon reduction and emission reduction [2]. Anaerobic ammonia oxidation has the advantages of no additional carbon source; no aeration; no additional alkalinity; and low sludge production [3]. Anaerobic ammonia-oxidizing bacteria (AAOB) directly converts ammonia (NH_4^+-N) and nitrite (NO_2^--N) into nitrogen and small amounts of nitrate (NO_3^--N) in an anaerobic environment and without the involvement of an organic carbon source [4]. However, in conventional municipal wastewater, which contains almost no nitroso-nitrogen (NO_2^--N), stable nitrosation is required to oxidise some of the ammonia nitrogen (NH_4^+-N) in the wastewater to nitroso-nitrogen (NO_2^--N) to provide a reaction substrate for anaerobic ammonia oxidation. Therefore, how to accumulate nitrite nitrogen (NO_2^--N) and enrich ammonia oxidising bacteria (AOB) to inhibit the activity of nitrite oxidising bacteria (NOB) is currently a challenge in the engineering of mainstream anaerobic ammonia oxidation of municipal wastewater [5].

To achieve stable partial nitrosation, it has been shown that: by inhibiting the activity of nitrite oxidising bacteria (NOB) in the system, which can usually be achieved by controlling operating parameters such as pH [6], aeration method [7], DO [8], free ammonia (FA) and free nitrite (FNA) [9] in the nitrosation reactor. According to relevant studies, zeolites have been reported to have strong selective adsorption properties for ammonia nitrogen (NH4 + -N). Based on the ability of zeolites to adsorb ammonia nitrogen and to convert to biodesorption after hanging membranes [10], this results in the suppression of NOB activity in the reactor and the stable acquisition of nitrite to provide a reaction substrate for subsequent anaerobic ammonia oxidation.

Stable partial nitrosation is a prerequisite for the stable operation of the integrated Partial Nitrification - Anammox (PN/A) process or Simultaneous Nitritation, Anammox and Denitrification (SNAD) [] process, the core of which lies in achieving enrichment of AOB and effective inhibition of NOB, ultimately leading to nitrite accumulation. A study by Jun Cheng [11] showed that the NAR (nitrite accumulation rate) of the reactor operating at a low ammonia nitrogen influent load was lower than that operating at a high ammonia nitrogen influent load, and therefore nitrite accumulation could not be achieved. Therefore, ammonia nitrogen loading is a key factor influencing the stabilisation of partial nitrosation. In practice, fluctuations in influent ammonia load are a common problem, and when changes in ammonia load lead to disruption of partial nitrosation, this leads to an increase in the ratio of effluent NH_4^+-N to NO_2^--N. The key to the successful operation of stable partial nitrosation is that the ratio of NH_4^+-N to NO_2^--N in the effluent meets the required influent requirements for the anaerobic ammonia oxidation process. Therefore, it is particularly important to quickly restore the reactor to stable operation and meet the requirements of the subsequent anaerobic ammonia oxidation process after a change in influent ammonia and nitrogen load.

In this study, an up-flow sludge fixed bed was used as the reaction vessel and natural zeolites with a particle size of 2–4 mm were added to the bottom of the reactor to investigate the activation of stable partial nitrosation under the combined effect of gradually increasing ammonia load and low DO conditions and the recovery of the system when the influent ammonia load fluctuated.

2 Materials and Methods

2.1 Experimental Procedures and Reaction Apparatus

The device used for the experiment is shown in Fig. 1. The whole device is made of glass with an effective volume of 2.5 L and is filled with 50 g/L of natural zeolite with a particle size of 2–4 mm. The bottom of the reactor is equipped with an aeration device, with a gas flow meter controlling the aeration volume to provide dissolved oxygen for the aerobic microorganisms in the system, and the bottom is equipped with a stirring device to prevent the separation of mud and water. The temperature inside the reactor is room temperature (26 ± 2) °C.

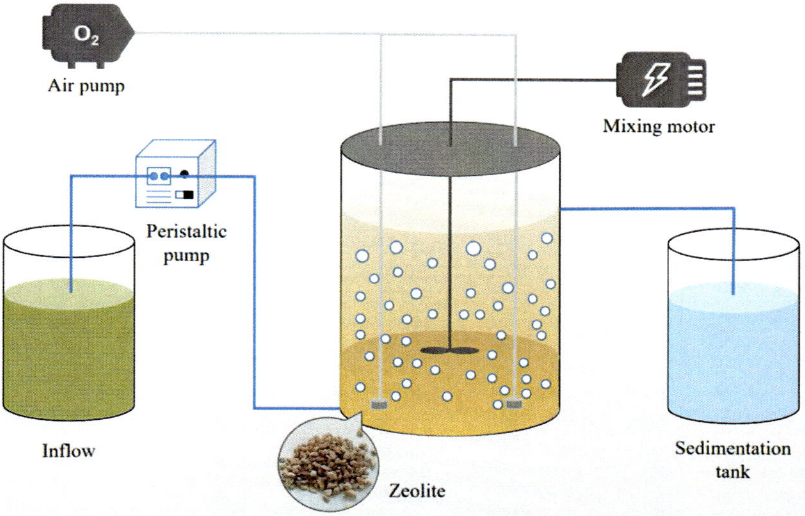

Fig. 1. Reactor device diagram

DO of the reactor in this study was chosen to be in the range of 1 ~ 1.5 mg/L [13 ~ 15]. Zongyue Yang [15] found that an aeration-stop ratio of 1:1 could stabilise NAR from 18% to over 90% after 22 d. pH is the controlling factor for stable nitrosation, and AOB growth is favoured within pH = 7.9 ~ 8.2 [6]. The experiments were carried out in two different stages and the operating conditions of the different experimental stages are shown in Table 1. Municipal wastewater.

Table 1. Operational conditions of the partial nitritation

Stage	Time	Feedwater ammonia nitrogen/(mg/L)	ALR/(kg/(m^3 d))	HRT (h)	Alkalinity (mg $CaCO_3$/L)
1	1-14d	130–300	0.13 - 0.3	24	650–1500
2	15-22d	300–350	0.45 - 0.525	16	1500–1750
3	23-37d	250	0.375	16	1250
4	38-45d	350	0.525	16	1750

2.2 Test Water and Inoculated Sludge

The experimental influent was artificially prepared simulated wastewater, in which ammonia nitrogen was provided by NH_4Cl, alkalinity by $NaHCO_3$, phosphorus source by KH_2PO_4 and no organic carbon source was added to the influent. The inoculated sludge was taken from the activated sludge of the secondary sedimentation tank of Henglin Wastewater Treatment Plant. Ruixin Wang [16] found that when the influent ammonia nitrogen was 500 mg/L, and the system was fed with alkalinity (in terms of $CaCO_3$) 2500 mg/L, i.e. ammonia nitrogen:alkalinity was 1:5, the best nitrosation effect was achieved with an average ammonia nitrogen conversion rate of 66.7%, an average nitrite nitrogen accumulation rate of 98.1% and an average nitrite nitrogen yield of 0.74 kg/(m^3 d).In this study, the influent ammonia nitrogen: alkalinity (as $CaCO_3$) = 1:5, with NH_4CI ranging from 573.3 to 1337.7 mg/L, $NaHCO_3$ from 630 to 1470 mg/L, KH_2PO_4 from 136 mg/L, $CaCO_3$ from 136 mg/L and $MgSO_4$-$7H_2O$ was 20 mg/L; trace element solution 1 mL/L. The trace element composition is shown in Table 2.

Table 2. Composition of trace elements

Trace elements	Mass concentration/(mg/L)	Trace elements	Mass concentration/(mg/L)
EDTA	10	$FeCl_3$-$6H_2O$	1.50
KI	0.18	$CoCl_2$-$6H_2O$	0.15
H_3BO_3	0.15	$MnCl_2$-$4H_2O$	0.12
$ZnSO_4$-$7H_2$ O	0.12	Na_2MoO_4-$2H_2O$	0.06
$CuSO_4$-$5H_2O$	0.12		

Note: All reagents are AR grade

2.3 Analysis Methods

Ammonia nitrogen (NH_4^+-N), nitrate (NO_3^--N) and nitrite (NO_2^--N) were determined by nano reagent spectrophotometry, UV spectrophotometry and N-(1-naphthyl) - ethylenediamine spectrophotometry respectively; pH was determined by PHBJ-260

(Remagnetics, Shanghai, China) and DO was determined by JPB-607A (Remagnetics, Shanghai, China). The design parameters for this study were calculated as shown in Eq. (1) to Eq. (7).

$$ARE = \frac{\left[NH_4{}^{+} - N\right]_{inf} - \left[NH_4{}^{+} - N\right]_{eff}}{\left[NH_4{}^{+} - N\right]_{inf}} \times 100\% \tag{1}$$

$$NAR = \frac{\left[NO_2{}^{-} - N\right]_{eff}}{\left[NO_2{}^{-} - N\right]_{eff} + \left[NO_3{}^{-} - N\right]_{eff}} \times 100\% \tag{2}$$

$$ALR = \frac{[NH_4^{+} - N]_{inf} \times Q}{V \times 1000} \tag{3}$$

$$V_1 = \frac{[NO_2^{-} - N]_{eff} \times Q}{V \times 1000} \tag{4}$$

$$V_2 = \frac{[NO_3^{-} - N]_{eff} \times Q}{V \times 1000} \tag{5}$$

$$\eta_1 = \frac{[NO_2^{-} - N]_{eff}}{\left[NH_4{}^{+} - N\right]_{inf}} \times 100\% \tag{6}$$

$$\eta_2 = \frac{[NO_3^{-} - N]_{eff}}{\left[NH_4{}^{+} - N\right]_{inf}} \times 100\% \tag{7}$$

where ARE is the ammonia removal efficiency, %; NAR is the nitrite accumulation rate, %; ALR is the influent ammonia load, kg/(m^3 d); $[NH_4{}^{+}\text{-}N]_{inf}$ is the influent ammonia concentration, mg/L; $[NH_4{}^{+}\text{-}N]_{eff}$ is the effluent ammonia concentration, mg/L; $[NO_3{}^{-}\text{-}N]_{eff}$ is the effluent nitrate concentration, mg/L; $[NO_2{}^{-}\text{-}N]_{eff}$ is the effluent nitrate concentration, mg/L 1; $[NO_2{}^{-}\text{-}N]_{eff}$ is the effluent nitrate concentration, mg/L; V_1 is the volumetric nitrosation rate, kg/(m^3 d); V_2 is the volumetric nitrification rate, kg/(m^3 d); η_1 is the nitrosation efficiency, %; η_2 is the nitrification efficiency, %; V is the volume of the reactor, L; Q is the influent flow rate, L/d.

3 Experimental Results and Analysis

3.1 Start-Up Phase

In this experiment, intermittent aeration was used to alternate the aerobic/anoxic environment in the reactor and to gradually increase the influent ammonia and nitrogen concentration to achieve NOB suppression. Therefore, during the nitrosation start-up phase (1–22 d), in order to effectively enrich AOB to suppress NOB, the HRT was kept at 24 h for the first 15 d. The experimental influent ammonia nitrogen concentration was gradually increased from 130 mg/L to 300 mg/L, and the influent ALR was also increased from 0.13 kg/(m^3 d) to 0.3 kg/(m^3 d); for 15–22 d, the HRT was shortened The

HRT was 16 h, while the influent ammonia nitrogen concentration was further increased to 350 mg/L, resulting in a final maximum ALR of 0.525 kg/(m^3 d) and a maximum effluent NPR (nitrite yield) of 0.24 kg/(m^3 d). The reactor started a partial nitrosation system for a total of 24 d. The influent ammonia nitrogen in the reactor, effluent ammonia nitrogen, nitrate, and nitrite concentrations are shown in Fig. 2.

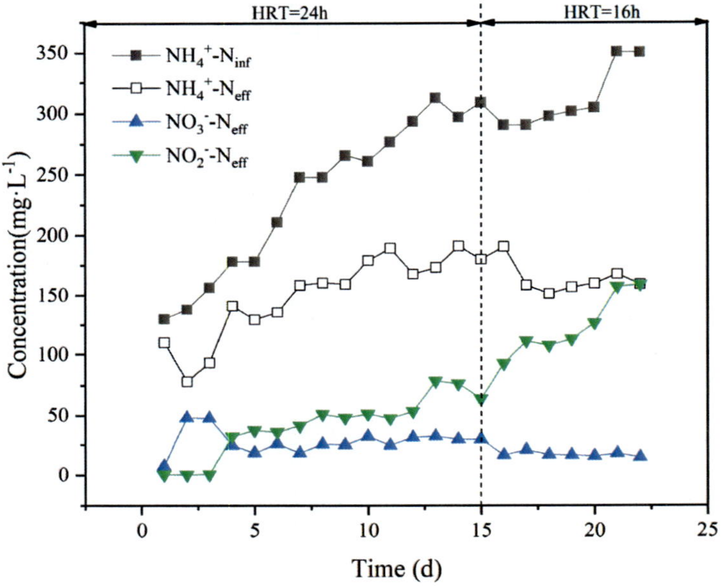

Fig. 2. Reactor nitrogen conversion

During the start-up phase, the influent ammonia nitrogen concentration was gradually increased from 130 mg/L to 350 mg/L. By day 4 of the experiment, the influent ammonia nitrogen concentration increased from 130 mg/L to 178.2 mg/L, as shown in Fig. 3, and the influent ammonia nitrogen load also increased from 0.13 kg/(m^3 d) to 0.18 kg/(m^3 d). Figure 4 shows that the nitrosation efficiency in the system at this time was 17.98% and the effluent nitrite concentration was 32.03 mg/L, which was higher than the effluent nitrate (24.63 mg/L), as shown in Fig. 3, the NAR was higher than 50% and the NO_2^--N accumulation rate was greater than 50% which is a sign of nitrosation and the system can be considered to have achieved nitrosation at this time. However, as the reactor DO was controlled at 1 ~ 1.5 mg/L, resulting in an ARE of only 27.44%, the effluent could not meet the influent requirements of the subsequent anaerobic ammonia oxidation process, so the influent ammonia N concentration was continued to be increased; and to further increase the influent ammonia N load, the HRT was shortened from 24h to 16h on day 15 and the ALR was increased from 0.3 kg/(m^3 d) to 0.45 kg/(m^3 d); then from day 17, the effluent nitrite concentration was 110.9 mg/L and the effluent nitrate concentration decreased to 20.8 mg/L, after which the NO_2^--N accumulation rate in the reactor was always maintained above 80% and the ARE also reached near 50%, indicating that the reactor had achieved stable partial nitrosation, at which time the system By day 21, the influent ammonia N was raised to 350 mg/L and the ALR was increased to 0.525 kg/(m^3 d). The effluent nitrite concentration reached 157.17 mg/L and the volumetric nitrosation rate was 0.2358 kg/(m^3 d), while the effluent nitrate remained the effluent NH_4^+-N/NO_2^--N $\approx$1, indicating that partial nitrosation was successfully initiated in the reactor and that the effluent could meet the influent requirements of the subsequent anaerobic ammonia oxidation process.

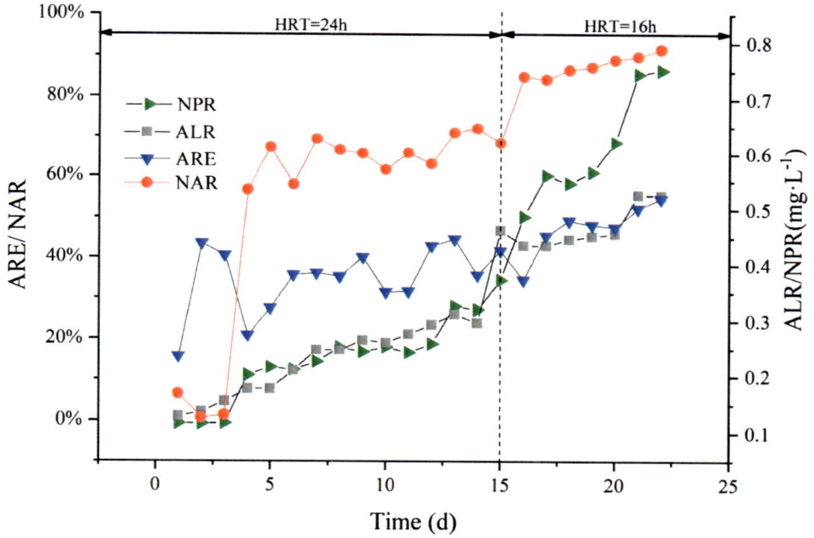

Fig. 3. Reactor nitrogen load changes

3.2 Reactor Imbalance and Recovery

When the influent ammonia nitrogen in the reactor was reduced, Jun Cheng [11] used the method of keeping the load constant and restoring the normal influent concentration, which failed to restore the denitrification performance of the system. Therefore, changing only the ammonia nitrogen under normal conditions could not restore the denitrification efficiency of the reactor. In contrast, CAPODICI [17] achieved stable operation of partial nitrification at an ammonia load of 0.30–0.50 kg/(m^3 d) and a C/N of 2–4 using a combined influent ammonia load and C/N, and this method required the addition of a carbon source when treating wastewater with low C/N. In addition, the use of increased aeration time [18] and control of DO also enhances the shock tolerance of partial nitrosation under high ammonia-nitrogen feed water and achieves stable partial nitrification process initiation. In this study, the bottom of the reactor was filled with 50 g/L natural zeolite. With the adsorption and desorption capacity of zeolite for ammonia nitrogen, when treating high concentration ammonia nitrogen wastewater, some of the ammonia nitrogen in the wastewater was adsorbed by the zeolite and the Z-Na^+ in the zeolite was

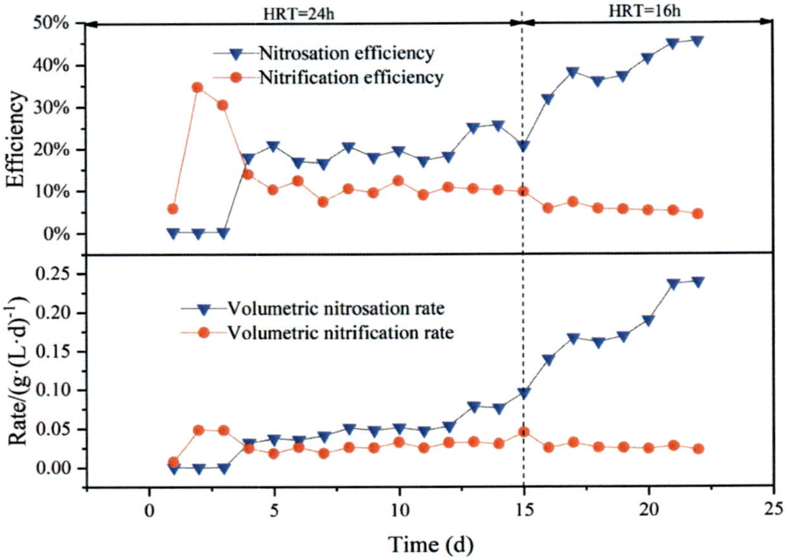

Fig. 4. Volumetric nitrosation efficiency/nitrification efficiency and volumetric nitrosation rate/nitrification rate

transformed into Z-NH_4^+. Meanwhile, when the ammonia nitrogen in the water was oxidized to nitrite by ammonia oxidizing bacteria, the Z-NH_4^+ in the zeolite was transformed into Z-Na^+ again, which transformed the zeolite adsorbed This maintains the FA concentration in the system by releasing the ammonia nitrogen. This process reduces the shock to the entire microbiological system caused by changes in influent ammonia nitrogen concentration and speeds up the recovery of some of the partical nitrosation when the system is damaged. The reactor imbalance and recovery took place over a total of 23 d. The influent ammonia nitrogen, effluent ammonia nitrogen, nitrate and nitrite concentrations in the reactor during this phase are shown in Fig. 5.

According to Fig. 6, at d 23, the influent ammonia nitrogen was reduced from 350 mg/L to 250 mg/L, so the ARE dropped sharply to 40% due to the sudden reduction in influent ammonia nitrogen, and the NOB in the system began to slowly regain activity; at d 28, the effluent nitrite concentration dropped to 20.18 mg/L, at which point the NAR dropped below 50%, at which point part of the nitrosation system was destroyed and The inhibitory effect of high ammonia load on NOB was reduced and some nitrite was oxidised to nitrate, maintaining this influent ammonia concentration, and by 37 d the effluent nitrate concentration and volumetric nitrification rate reached maximum values of 91.49 mg/L and 0.1374 kg/(m^3 d), respectively, at which point the effluent nitrite concentration and volumetric nitrosation rate were 24.83 mg/L and 0.0373 kg/(m^3 d) respectively, with the NAR reduced to 21.35% as obtained from Fig. 7. Subsequently, the influent ammonia concentration of 350 mg/L was restored again from 38 d to see if the system could recover its nitrosation performance and it was observed that at 40 d, the NAR increased rapidly to 71.22%, the effluent nitrite concentration and volumetric

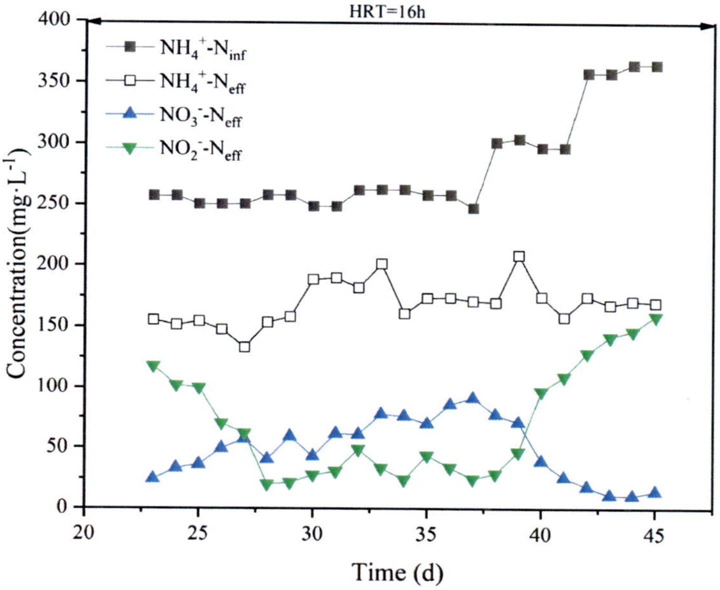

Fig. 5. Reactor nitrogen conversion

nitrosation rate were 97.33 mg/L and 0.146 kg/(m^3 d) and the effluent nitrate concentration and volumetric nitrification rate were 39.33 mg/L and 0.0590 kg/(m^3 d); the effluent NAR was stable above 90% and effluent NH_4^+-N/NO_2^--N ≈ 1 when the influent ammonia nitrogen concentration was continuously maintained until 43 d, indicating that good nitrosation could be quickly restored in the reactor and the effluent conditions were met. This is similar to the findings of Ping Zhong [19], where in a continuous flow zeolite SBR reactor, when the system was disrupted, it took only 4 d to restore the ammonia nitrogen to normal levels to restore stable partial nitrosation. The addition of zeolites in the reactor not only provided microorganisms with attachment sites and space for growth, but also allowed for proper regulation of FA concentrations in the system based on the adsorption and desorption of ammonia nitrogen by zeolites when the influent water received fluctuations, and the good nitrosation performance could be quickly restored after restoring the influent ammonia nitrogen concentration when the system was damaged.

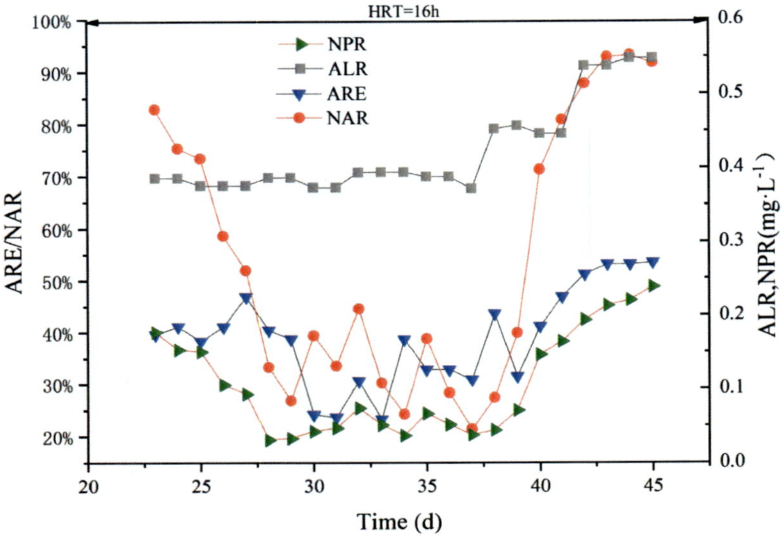

Fig. 6. Reactor nitrogen load changes

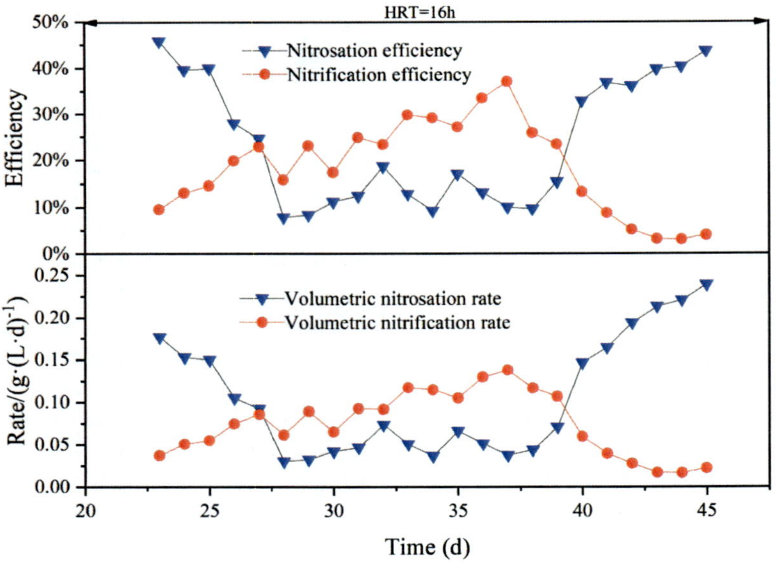

Fig. 7. Volumetric nitrosation efficiency/nitrification efficiency and volumetric nitrosation rate/nitrification rate

4 Conclusion

Stable partial nitrosation was successfully achieved in the upflow sludge fixed bed reactor after 21 d at ambient temperature (26 ± 2) °C by gradually increasing the influent ammonia nitrogen and shortening the HRT, with the effluent meeting the requirements of the subsequent anaerobic ammonia oxidation influent.

Upflow sludge fixed bed stabilised partial nitrosation performance: In the stabilisation phase, the influent ammonia nitrogen load was 0.525 kg/(m^3 d), the NAR was above 90%, the ARE was around 53% and the maximum volumetric nitrosation rate was 0.454 kg/(m^3 d), Effluent NH_4^+-N/NO_2^--N≈1.

When the reactor influent ammonia nitrogen concentration was reduced, the system was disrupted and the NAR dropped to as low as 21.35%. When the influent ammonia nitrogen concentration returned to normal, the NAR recovered to over 80% after only 4 d; at d 6, the NAR was maintained at over 90%. When the water ingress fluctuates, the reactor can quickly restore normal nitrosation performance, and the system is more stable in terms of impact response in practical engineering applications.

Acknowledgments. This work was funded by Fujian provincial industry-universityresearch collaborative innovation (2021Y4005), Fujian Science and Technology Guiding Project (2020Y0056).

References

1. OlavSliekersa, A., Derworta, N., CamposGomezb, J.L., et al.: Completely autotrophic nitrogen removal over nitrite in one single reactor. Water Res. **36**(10), 2475–2482 (2002). https://doi.org/10.1016/S0043-1354(01)00476-6
2. Tang, J., Xie, J., Chen, Z., et al.: Carbon neutral technology and case study of urban wastewater treatment plant. Chem. Ind. Eng. Prog. **41**(05), 2662–2671 (2022)
3. Ji, J., Peng, Y., Li, X., et al.: A novel partial nitrification-synchronous anammox and endogenous partial denitrification (PN-SAEPD) process for advanced nitrogen removal from municipal wastewater at ambient temperatures. Water Res. 175 (2020)
4. Cao, Y., van Loosdrecht, M.C.M., Daigger, G.T.: Mainstream partial nitritation–anammox in municipal wastewater treatment: status, bottlenecks, and further studies. Appl. Microbiol. Biotechnol. **101**(4), 1365–1383 (2017). https://doi.org/10.1007/s00253-016-8058-7
5. Gu, J., Yang, Q., Liu, Y.: Mainstream anammox in a novel A-2B process for energy-efficient municipal wastewater treatment with minimized sludge production. Water Res. **138**, 1–6 (2018)
6. Reino, C., Suárez-Ojeda, M.E., Pérez, J., Carrera, J.: Stable long-term operation of an upflow anammox sludge bed reactor at mainstream conditions. Water Res. **128**, 331–340 (2018). https://doi.org/10.1016/j.watres.2017.10.058
7. Li, N., XiaoMin, H., Li, G., et al.: Effect of HRT and pH on short-range nitrification and denitrification in MBBR. Ind. Water Treat. **10**(36), 20–23 (2016)
8. Qian, R., Jiang, H., Liu, S., et al.: Carbon capture pretreatment of black water based on controlled aeration method for partial nitrosation nitrification process. Chinese J. Environ. Eng. **04**(16), 1353–1362 (2022)
9. Blackburne, R., Yuan, Z., Keller, J.: Partial nitrification to nitrite using low dissolved oxygen concentration as the main selection factor. Biodegradation **19**(2), 303–312 (2008). https://doi.org/10.1007/s10532-007-9136-4
10. Yao, Q., Peng, D., Wang, B., et al.: Effect of free ammonium and free nitrous acid on the activity, aggregate morphology and EPS distribution of ammonium oxidizing bacteria in partial nitrification. J. Biosci. Bioeng. **124**(3), 319–326 (2017)

11. Chen, Z., Wanga, X., Chen, X., et al.: Nitrogen removal via nitritation pathway for low-strength ammonium wastewater by adsorption, biological desorption and denitrification. Biores. Technol. **267**, 541–549 (2018)
12. Chen, J., Zhang, L., Zhang, S., et al.: Influence of ammonia nitrogen load fluctuations on the nitrate state of short-course nitrification-anaerobic ammonia oxidation process in municipal wastewater nitrogen. China Environ. Sci. **37**(02), 520–525 (2017)
13. Wang, H., KunMing, F., Zuo, Z., et al.: Effect of hydraulic retention time and dissolved oxygen on vitrified CANON reactors. Environ. Sci. **36**(11), 4141–4147 (2015)
14. Xue, Y., Yang, F., Liu, S., et al.: The influence of controlling factors on the start-up and operation for partial nitrification in membrane bioreactor. Biores. Technol. **100**(3), 1055–1060 (2009)
15. Tian, W.-D., An, K.-J., Ma, C., et al.: Partial nitritation for subsequent anammox to treat high-ammonium leachate. Environ. Technol. **34**(8), 1063–1068 (2013)
16. Yang, Z., KunMing, F., Liao, M., et al.: Exploration of inhibition strategies of 2 nitrite oxidizing bacteria during short-range nitrification. Chin. J. Environ. Eng. **13**(1), 222–231 (2019)
17. Wang, R., Chen, J., Wang, X., et al.: Effect of alkalinity on nitrosation in zeolite sequencing batch reactors. Environ. Sci. **40**(6), 2807–2812 (2019)
18. Capodici, M., Corsino, S.F., Di Trapani, D., et al.: Achievement of partial nitrification under different carbon-to-nitrogen ratio and ammonia loading rate for the co-treatment of landfill leachate with municipal wastewater. Biochem. Eng. J. 149 (2019)
19. Okabe, S., Oshiki, M., Takahashi, Y., et al.: Development of long-term stable partial nitrification and subsequent anammox process. Biores. Technol. **102**(13), 6801–6807 (2011)
20. Zhong, P., Wang, X., Qin, J., et al.: Continuous flow zeolite SBR for stable nitrosation of high ammonia nitrogen wastewater. Technol. Water Treat. **48**(2), 93–97 (2022)

Immobilized Anaerobic Digestion of Molasses-Based Distillery Wastewater in Moving Bed Biofilm Reactor

Michelle Almendrala[1(✉)], Shekinah Mae Villaflor[1], Zhane Ann Tizon[1], Bonifacio Doma[1], and Ralph Carlo Evidente[2]

[1] Mapúa University, 1002 Manila, Philippines
mcalmendrala@mapua.edu.ph
[2] POSTECH, Pohang, Gyeongbuk 37673, South Korea

Abstract. Anaerobic digestion of ethanol-based distillery wastewater (DWW) was conducted in an immobilized 2-L Erlenmeyer flask set in 120 rpm incubator shaker and at 40°C thermophilic condition for 30 days. K1 moving bed biofilm carriers, with specifications of 10 mm diameter, 7 mm length, density of 0.96–0.98 g/cm^3 and a specific biofilm surface area of > 800 m^2/m^3, were used for immobilization. With five batches of anaerobic digestion, pure and diluted DWW were observed. The control batch with no carriers and with pure DWW produced the least volume of biogas. Batch 5, whose volume dilution ratio is 7:3 (DWW: pure water), produced the highest volumetric yield of biogas of 2 L. Similarly, Batch 2, which is pure DWW with no plastic carriers produced almost the same volume of biogas (1850 L) as compared to that of Batches 3 through 5 with complete nutrients. Among the samples, Batch 5 has the highest degree of TSS reduction and has BOD/COD ratio of 0.23, which is closest to zero that practically implies perfect biodegradability.

Keywords: DWW · Anaerobic digestion · Moving bed biofilm · Immobilization

1 Introduction

Solid Organic Waste (SOW) management still faces many challenges in developing countries, with food wastes, yard wastes, animal manure, waste activated sludge, and agricultural waste yield in an increasing annual rate. One of the most polluting organic wastes is wastewater from distilleries due to the 88% of its raw materials that is discharged into the environment [1]. Ethanol as a type of alcohol serves many purposes in various industries, such as in chemical and manufacturing. Ethanol industry is producing huge amounts of alcohol to supply to the society. However, large volume of this unwanted chemical liquid is emitted and posed a threat to the ecosystem as it contains high biological and chemical oxygen demand, salt contents, and heavy metals that makes it acidic [2]. There exist many methods to resolve this global issue by converting the waste into valuable products. The biofuel as a research area gives a promising solution in producing energy from these SOW, and simultaneously utilizing the remaining waste

X. Chen (Ed.): ICESE 2022, ESE, pp. 29–36, 2023.

as a possible resource or raw material to a significant industry. With potential in treating various organic wastes, anaerobic digestion (AD) is used to transform these into biogas energy and a stabilized fertilizer known as digestate. However, limitations in anaerobic digestion originate from the inadequate amount of mono-substrate supply and lack of immobilization of the feedstock. For the AD to become an efficient process, there must be an optimal amount of nutrients such as carbon (C), nitrogen (N), phosphorus (P), and sulfur (S). However, if the substrate has an excessive or deficient C/N, it can cause incomplete conversion of carbon due to inability in producing enzymes or can accumulate ammonia from conversion of urea [3].

Microbial culture present during anaerobic digestion should be immobilized to maintain a low concentration of the intermediates. A moving bed biofilm reactor has proven its potential in wastewater treatment to reduce the harmful organic compounds present. In terms of cost effectiveness in maintaining a high rate AD, the need to use moving bed biofilm reactor system sustains the microbiological community by the growth of biofilm in the plastic media. To this date, no studies found in the previous researches in utilizing a moving bed biofilm reactor.

This study aims to achieve the following objectives: (1) to compare the methane yield of distillery wastewater digested under plastic media to that of recent studies and (2) to determine the physico-chemical characteristics of the final digestate: pH, intial and final BOD/COD, %TSS (Total Suspended Solids) reduction, and %TP (Total Phosphorus). This papers seeks to analyze the effects of biofilm formation in the improvement of biogas supply using a moving bed biofilm reactor. Further, this can provide new insights in determining the effects of added nutrients and other factors that may influence the biogas production.

2 Experimental Methodology

Distillery wastewater (DWW) samples were freshly taken from Batangas' Absolut Distillers, Inc, whereas the cow manure was sourced from YGGACC HAI farms in Laguna. The latter was sealed and incubated at 40 °C anaerobically for seven (7) days before the experiment. Cow manure is used as the inoculum for the samples to generate bacteria immediately throughout the experiment. In this study, there are five samples containing distillery wastewater and pure water. Table 1 shows different parameters for each batch sample, where the DWW/water ratio, micronutrients, glucose, and media carrier are taken into consideration. The composition of the micronutrients used can be found in Table 2.

In order to immobilize the substrates, plastic carriers are used as the medium to contain the bacteria for anaerobic digesting upon integrating in the moving bed biofilm reactor. K1 moving bed biofilm carriers that were utilized has the following specifications of 10 mm diameter, 7 mm length, density of 0.96–0.98 g/cm^3 and a specific biofilm surface area of $> 800\ m^2/m^3$.

To remove the oxygen and promote AD in all the samples, nitrogen gas entered through a glass tube, submerged in the mixture, and was purged in each of the media bottle for 10 min. The oxygen that was initially present in the samples exit through the other glass tube. In order to collect the produced biogas, 2-L urine bags were connected

Table 1. Different controls used for the anaerobic digestion study

Sample Batch	DWW (mL)	H_2O (mL)	Nutrient	Glucose (g/200 mL)	Carriers
1	1400	0	X	X	X
2	1400	0	✓	✓	X
3	1400	0	✓	✓	✓
4	980	420	X	X	✓
5	980	420	✓	✓	✓

to each of the media bottles though rubber tubes. The media bottles were exposed to a constant temperature of 37°C inside an orbital shaker set at 120 rotations per minute in a duration of 4 weeks as shown in Fig. 1.

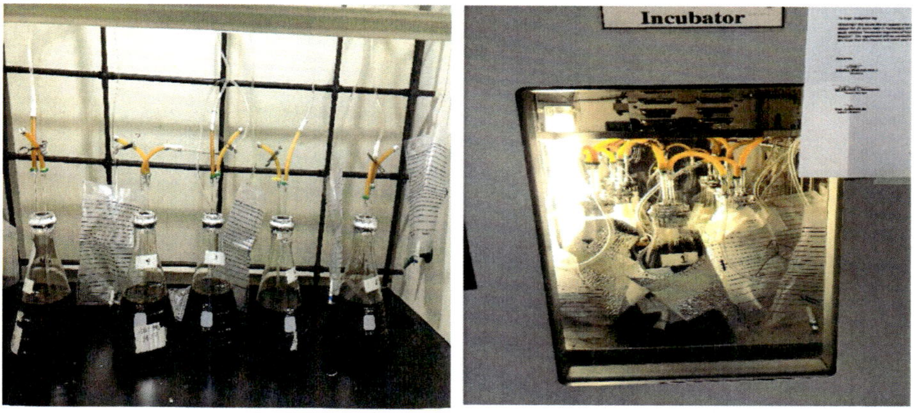

(a) (b)

Fig. 1. Experimental setup for lab-scale Anaerobic Digestion: (top) Sample Batches and (bottom) thermophilic condition imposed by thermal incubator

Table 2. List of Macronutrients and Micronutrients used for the anaerobic digestion study

Nutrients	Required Amt. Based on the Literature (g/L) [2]	Weighed Amount (g)
Glucose	30	42.0364
KH_2PO_4	0.5	0.7033
$MgSO_4.7H_2O$	0.2	0.2863
$MnSO4.7H_2O$	0.01	0.0163

(continued)

Table 2. (*continued*)

Nutrients	Required Amt. Based on the Literature (g/L) [2]	Weighed Amount (g)
$FeSO_4.7H_2O$	0.01	0.0146
NaCl	0.01	0.0152
$CuSO_4.5H_2O$	0.1	0.1437
$CoCl_2.6H_2O$	0.1	0.1444
$ZnSO_4.7H_2O$	0.02	0.0317

3 Results and Discussion

Figure 2 shows that using pure DWW produced the least biogas as the authors speculated less amount of carbon utilized by the anaerobes. Batches 2 and 5, meanwhile, produced the maximum volume of around two (2) L at day = 24; however, the former diffused from 2000 mL to 1850 mL at the last day. Both of them had nutrients and glucose that suit as optimal intake for the bacteria, which led the AD process to commence within a short period of time.

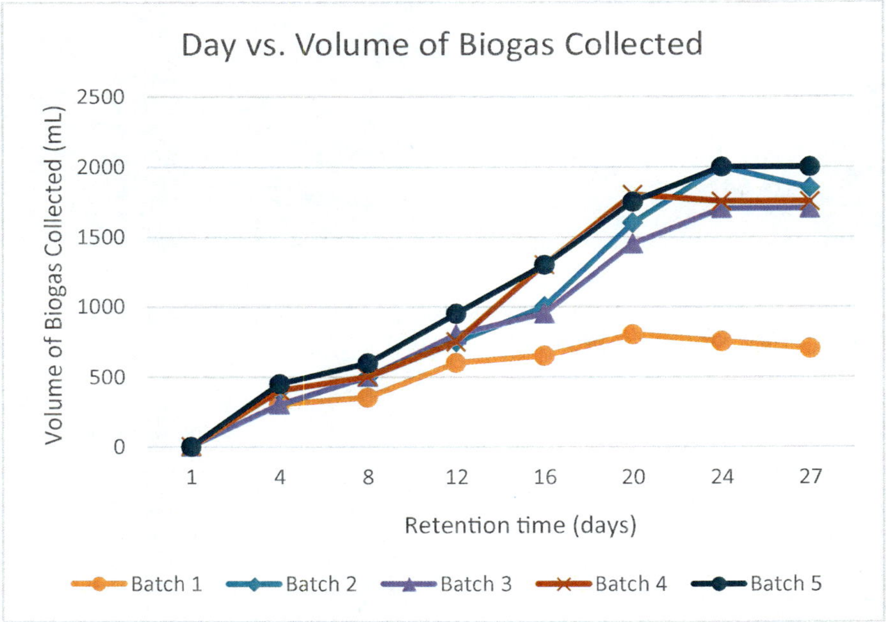

Fig. 2. Volumetric profile of Biogas for 27-day period

During the digestion process, there is a decrease in pH for all of the batch samples as there is a production of acetate and fatty acids [4]. The methanogenic bacteria can halt the biological reactions in the digestion due to its sensitivity in a very low pH environment. The optimal pH of final digestate ranges from 4.0 to 7.2 in anaerobic digestion [5].

In Fig. 3, Sample 4, which has the lowest decrease of pH, resulted to low biogas yield, whereas the remaining batch samples gave a consistent change of pH. The BOD/COD ratio in percentage depends upon the biodegradability of organic matter, which ranges from 0% (completely non-biodegradable) to 100% (completely biodegradable). Generally, it decreases before to after digestion (AD) due to presence of high proportion of non-degradable materials [5]. The optimal BOD/COD before undergoing to AD spans a range from 0.39 to 0.55, based on literature [6]. Generally, data points show that biodegradation took the most effect on the sample of diluted distillery wastewater mixed with nutrients, glucose, and media, thereby providing a place for colonies of bacteria to grow. Data of the change in BOD to COD ratio are well-matched with the AD principle by showing the largest difference between the initial and the final values. The BOD/COD ratio in Sample 5 is approaching to zero implying that degradation took place in an effective manner, with the possibility of projecting a hitting peak state wherein biodegradation stopped.

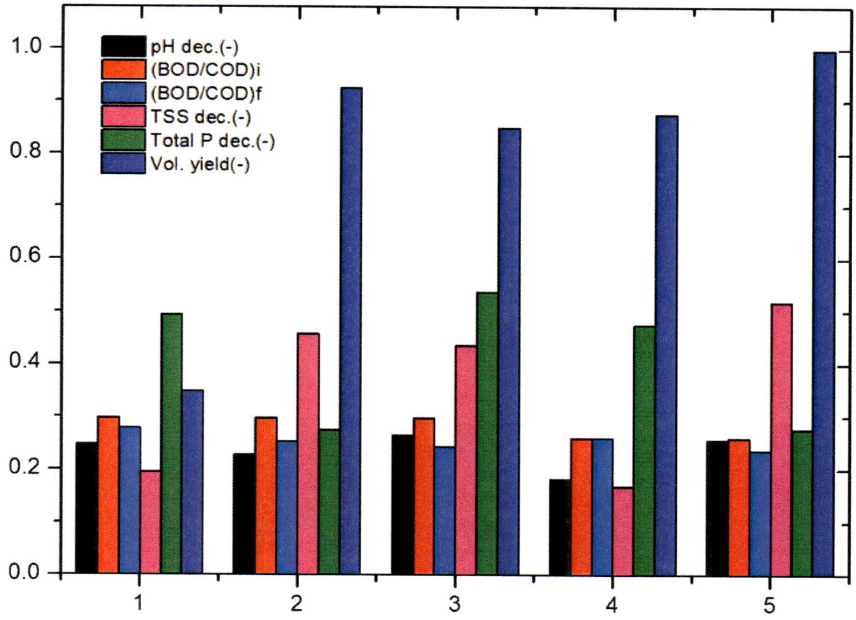

Fig. 3. Comparison of different parameters among all Batch Samples

Also, Sample 5 showed the greatest reduction of suspended solids, indicating that AD took the greatest effect on it by means of biodegradability that was also observed in BOD/COD ratio. This only indicates that higher % reduction value of TSS resulted to greater volume of biogas yield. In this study, the optimum TSS to generate greater volume of biogas should be in the range of 938 to 1,456 mg/L. On the other hand, the data for both biogas yield and total phosphorus shows no consistent trend and thus implies no correlation.

It is noteworthy that only Batch 5 was the only sample the researchers observed with the formation of biofilm as seen in Fig. 4. This phenomenon could be attributed to the enhanced metabolic processes due to coordinated interactions between individual microbes that exist in a single culture of mixed microbes. Applicable for Sample 5, the microbial cultures in the biofilm seemingly digest the nutrients such as phosphorous and nitrogen-containing complexes during AD.

(a) (b) (c)

Fig. 4. Comparison of media carriers after 27 days of AD: (a) Batch 3, (b) Batch 4, and (c) Batch 5

Table 3 summarizes the quantitative biogas yield from the initial volumetric amount of the substrates used. The volume of the feed is constant for all the samples, which is 1.4 L pure distillery wastewater. However, the biogas yield differs as there is an addition of nutrients, glucose, and carriers. The ratio of distillery wastewater and pure water also influenced the amount of biogas produced through anaerobic digestion. It is shown in this study that a mixture of distillery wastewater and pure water with added nutrients, glucose, and utilization of plastic carriers produced the highest yield among the other samples for about 2 L. It is also to highlight that with a mono-substrate of distillery wastewater with nutrients and glucose only generated the second highest volume of biogas of 1.85 L.

Table 3. Comparison of substrate to its volume of biogas produced

Substrate	Feed (L)	Biogas yield (L)
Pure DWW with nutrients and glucose	1.4	1.85
Pure DWW with nutrients, glucose, and carriers	1.4	1.7
49:21 (DWW to H_2O) with carriers	1.4	1.75
49:21 (DWW to H_2O) with nutrients, glucose, and carriers	1.4	2

4 Conclusion

In this batch experimental study, indicators such as pH level, BOD/COD ratio and total suspended solids are considered in the study on anaerobic digestion of sugarcane distillery wastewater and are compared from the volumetric biogas yield pH levels must be around the setpoint value of 5.0 for the sample to undergo anaerobic digestion. Potential candidates tend not to change pH drastically during AD due to the buffer capacity supplied by ion bicarbonate equilibrium that should exist in the sample. A ratio of BOD to COD approaching to zero practically implies perfect biodegradability. Among the samples, Sample 5 has the BOD/COD ratio closest to zero and is also the sample that yields the most biogas. Meanwhile, a reduction of suspended solids almost always resulted from the proceeding of anaerobic digestion.

This experimental study shows that a combination of cobalt, magnesium, iron, and zinc makes AD proceed faster. Another way for making anaerobic digestion more efficient is by employing packing materials, or a media to allow more interaction between the microbes and the substrate. It is also noted, that a blend of microbes is better than just having a culture of one kind of microbes. This is because a coordinated metabolic processing makes anaerobic digestion proceed better than just having one kind of microbes do the entirety of the metabolic processing. Although some parameters need to take account such as Volatile Suspended Solid (VSS) reduction, overall, this study gives better insights of digesting sugarcane distillery wastewater in an anaerobic environment with the aid of plastic media containing a culture of multiple types of microbes.

References

1. Kharayat, Y.: Distillery wastewater: bioremediation approaches. J. Integrative Environ. Sci. **9**(2), 69-91 (2012). https://doi.org/10.1080/1943815X.2012.688056
2. Ratna, S., Rastogi, S., Kumar, R.: Current trends for distillery wastewater management and its emerging applications for sustainable environment. J. Environ. Manage. **290**,(2021)
3. Gerardi, M.H.: The Microbiology of Anaerobic Digesters. John Wiley and Sons, Inc. Chapter 7, pp. 51–59 (2003)
4. Lindorfer, G., Ahring, B., Verstraete, W.: Pre-treatment technologies for enhanced energy and material recovery of agricultural and municipal organic wastes in anaerobic digestion. In: Proceedings on Future of Biogas in European Biogas Workshop, pp. 79–85 (2003)
5. Menon, A., Wang, J., Giannis, A.: Optimization of micronutrient supplement for enhancing biogas production from food waste in two-phase thermophilic anaerobic digestion. Waste Manage. **59**, 465–475 (2017)
6. Janke, L., Leite, A., Wedwitschka, H., Schmidt, T., Nikolausz, M., Stinner, W.: Biomethane production integrated to the brazilian sugarcane industry: the case study of São Paulo State. In: Proceedings of the 22nd European Biomass Conference and Exhibition, Hamburg, Germany, vol. 2326, p. 2326 (2014)

Effects of Straw Returning on Paddy Soil Environmental Remediation

Chunying Li[1,2] and Chunlong Zhang[2(✉)]

[1] School of Applied Meteorology, Nanjing University of Information Science and Technology, Nanjing, China
[2] West Anhui University, Lu'an, China
492346238@qq.com

Abstract. Straw is an important renewable resource for fertilizing soil, and returning straw to the field is the most economical and effective technical measure for fertilizing and improving soil. At the same time, it can solve the problem of air pollution caused by burning straw in the field. In this paper, the typical paddy soil in the northern subtropical region of the middle and lower reaches of the Yangtze River was taken as the research object, and the effects of different types of straw returning and fertilization on the carbon and nitrogen contents in paddy soil were analyzed. Rice and corn stalks were air-dried and ground, mixed with soil, chemical fertilizers and sufficient water, and incubated in a greenhouse at 40 °C for 16 days. The results showed that straw returning to the field could significantly increase soil organic matter content and improve soil fertility, and the organic matter content increased from 0.9% before the experiment to more than 3%. Straw returning to the field can increase the soil organic carbon content. In this experiment, the straw returning treatment increased the soil organic carbon content by more than 3 times on average compared with the no-straw returning treatment. At the same time, the order of fertilization and straw returning should be paid more attention to avoid the loss of nitrogen resulting in secondary pollution.

Keywords: Straw returning · Soil fertility · Total nitrogen · Organic carbond

1 Introduction

As a powerful measure to regulate soil nutrient cycle, reduce nitrogen fertilizer loss, and maintain and improve soil organic matter level, crop straw returning has been paid more and more attention by people. A large amount of straw is produced under the traditional agricultural planting system. Straw is a high-quality renewable organic resource [1], but the phenomenon of straw burning in production is still common [2], which not only wastes resources, but also harms the environment [3, 4]. Straw returning has many advantages such as increasing soil organic matter content, improving soil physical and chemical properties, increasing soil aggregates, promoting soil moisture and fertilizer conservation, and ensuring high corn yield [5–7]. Farmland management measures.

Predecessors have done a lot of work on the impact of straw returning to soil on carbon and nitrogen. These works mainly focused on the initial and final state of straw

X. Chen (Ed.): ICESE 2022, ESE, pp. 37–49, 2023.

returning to the field, and focused on the impact of the initial and final state of straw returning to the field on crop yields. There are few studies on the dynamic changes of soil carbon and nitrogen caused by the release of straw nutrients and the application of chemical fertilizers. It is generally believed that the C/N ratio of corn stalks is higher than that of rice stalks, and it is more difficult to decompose, and the nutrients in it cannot be released quickly for the current crops to use [8]. Due to the differences in the chemical composition of different plant straws, they will have different effects on soil pH after they decompose in the soil [9]. Returning straw to the field can significantly increase the soil microbial biomass, leading to an increase in soil available nitrogen (small straw C/N) or a decrease (large straw C/N). Chen Xingli et al. [10] conducted indoor cultivation experiments on corn stalks and showed that the application of chemical fertilizers or organic fertilizers can significantly increase the nitrogen content of the stalks and reduce the C/N ratio, which is conducive to the decomposition and release of the straw's own nutrients. Ding Xueli et al. [11] further research showed that within a certain range of nitrogen fertilizer application, the soil microbial biomass continued to increase with the increase of inorganic nitrogen application. This is mainly because straw usually contains a high C/N ratio, and a certain amount of active inorganic nitrogen needs to be applied from the outside to make up for the insufficient supply of active nitrogen in the soil during the straw decomposition process, and to meet the nitrogen absorption by the straw degrading microorganisms., To maintain a high number and activity of microorganisms, thereby increasing the rate of mineralization and decomposition of straw. Chen Shanghong et al. [12] showed that continuous straw returning to the field can increase the soil organic carbon content under the condition of two-cropping and upland rotation in a year, and the straw returning to the field can increase the soil organic carbon content more than no-tillage.

Returning straw to the field can increase soil nitrogen supply. The research of Hong Chunlai [13] showed that the total nitrogen in the paddy soil increased from 0.205% to 0.229% after returning the straw to the field, and the alkali hydrolyzed nitrogen increased by 25.5mg·kg-1. The results of the fertilization experiment in the micro-area of straw returning to the field in northern Liaoning showed that the returning of straw to the field significantly increased the soil available nitrogen, the treatment with the largest amount of returning to the field increased the available nitrogen by 11.18%, and the total N content was also significantly increased [14]. The results of Wang Shuping [15] showed that the addition of corn stalks and root stubble can significantly increase the nitrogen content of the soil. Compared with single application of chemical fertilizers, they increase by 0.096 g/kg and 0.094 g/kg, respectively. Li Xiaoyong et al. [16] analyzed the test results of returning straw to the field to improve soil fertility and showed that the total nitrogen content of the soil increased by 28.5% ~ 40.1%, and the content of alkali hydrolyzed nitrogen increased by 13.2% ~ 30.8%. Long-term application of chemical fertilizers, especially nitrogen fertilizers, can increase the total N and available N content in the soil. However, the applied inorganic nitrogen fertilizer seldom accumulates in the soil organic matter. Only when organic carbon (such as returning straw to the field) is increased at the same time, can the organic nitrogen content be increased, and its mineralization can be improved, which is conducive to biological absorption of nitrogen [17]. The combined application of straws and chemical fertilizers can promote the conversion between the mineralization and fixation of organic nitrogen in the soil, so that the distribution of various humus nitrogen in the soil and the C/N ratio are continuously coordinated and updated, and the nitrogen cycle in the soil-plant system is accelerated Role [18].

In this experiment, corn and rice straw were used as materials to study the effects of two treatments with and without fertilizer on the physical and chemical properties of the soil after returning the straw to the field under flooded culture conditions. The soil pH, electrical conductivity, organic matter and nitrogen content were measured for 16 days. The dynamic changes of soil carbon and nitrogen after returning straw to the field and fertilizing were discussed. In order to obtain the dynamic relationship between chemical fertilizers and the decomposition process of straw, the aim is to provide a scientific basis for the rational return of straw to the field and the full utilization of straw resources.

2 Materials and Methods

2.1 Soil Material

The paddy soil used in the experiment was taken from the Agricultural Meteorological Experiment Station of Nanjing University of Information Technology in November 2021 (32°03′N, 118°51′E). The local multi-year average temperature is 15.6°C, and the multi-year average precipitation is 1100 mm per year. The test soil was a stagnant paddy soil (Grey genus), the soil texture was loamy clay, and the clay content of the topsoil was 26.1%. The soil of the tillage layer was collected after the late rice harvest in 2011, and the sampling depth was 0–20 cm. The collected soil was air-dried, and the stones, rice roots, etc. were removed and stored for later use. The basic physical and chemical properties of the soil are as follows: soil bulk density 1.57, pH (water-soil ratio 2.5:1) 7.90, conductivity 0.35μs/cm, organic matter content 8.54g/kg, total nitrogen 739.66mg/kg, ammonium nitrogen 13.42mg/kg.

2.2 Straw Material

The straws used in the experiment were rice straws and corn straws collected after the crops were harvested in November 2021, which were air-dried and ground for later use.

2.3 Experiment Method

There are six treatments in this experiment: -FS (no fertilization and no straw), -F + M (no fertilization and corn stalk), -F + R (no fertilization and rice straw), + FS (no stalk fertilization), + F + M (fertilization plus corn stalks), + F + R (fertilization plus rice stalks). Fertilization treatment is 0.2gN/kg soil, calculated as $(NH_4)_2SO_4$; 0.15gP_2O_5/kg soil, calculated as $Ca(H_2PO_4)_2{\cdot}H_2O$; 0.15g KCl/kg soil; straw treatment is 5%, that is, every 100g soil add 5g straw. The rice and corn stalks are air-dried and ground and mixed with soil, chemical fertilizer and sufficient water. Weigh 200g of the mixed soil in a disposable plastic cup and cultivate in a greenhouse at 40°C for 16 days to measure soil pH, conductivity, organic matter, total nitrogen, ammonium nitrogen and nitrate nitrogen and other indicators.

Among them, the pH and conductivity are measured by electrode method (water-soil ratio is 2.5:1). The soil organic matter is determined by the potassium dichromate external heating method, the sample is digested with concentrated H_2SO_4 and titrated with $FeSO_4$; the total nitrogen in the soil is determined by the Kelvin method, the sample is added to the catalyst, the sample is digested with concentrated H_2SO_4, and the nitrogen analyzer is used After distillation, the distillate was titrated with H_2SO_4 (0.005M). Use the standard curve method (see Fig. 1) to calculate the results.

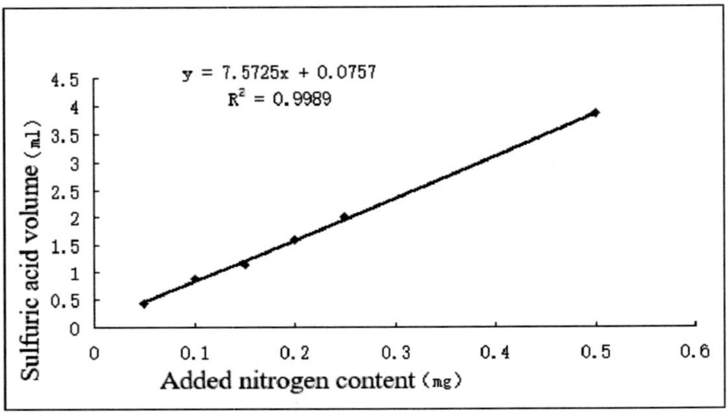

Fig. 1. Standard song for total nitrogen determination

2.4 Experiment Method

The experimental data using Excel 2007 basis calculated and analyzed using single factor analysis of variance (one-way ANOVA) differentially average significance test, $p < 0.05$ as significant difference, $p < 0.01$ for the difference was significant, the LSD method Multiple comparisons.

3 Results and Discussion

3.1 Effects of Straw Returning and Fertilization on Soil Structure

Returning straw to the field under crushed conditions can increase the total porosity of the soil and reduce the bulk density of the soil, and there will be no disadvantages such as the loss of soil nutrients due to the looseness of the soil and the uneven size of the soil voids. In this experiment, both corn stalks and rice stalks increased the porosity of the paddy soil. After the straw is crushed and added to the soil, the bulk density of the soil can be reduced. This may be because the crushed straw provides a better decomposition environment for soil microorganisms, which accelerates the decomposition rate of the straw, and its secretion products are beneficial to the soil aggregates. Formed, thereby improving the stability of the soil structure and improving the structural condition of the soil.

3.2 Changes of Soil pH After Returning Straw to the Field and Fertilizing

The changes of soil pH in the experiment are shown in Table 1. It can be seen from the table that the pH of the soil in the control group was 7.90 before the flooding treatment, and the soil was weakly alkaline. The soil pH decreased after adding straw and fertilizer, and the order of magnitude was CK > + F-S > + F + M > -F + M > -F + R > + F + R. In the process of culture, CK, + F + M, -F + M showed a trend of first decreasing (0-8d) and then increasing (8-16d). + F-S shows a trend of decreasing. -F + R and + F + R fluctuate slightly, first increase (0-4d), then decrease (4-8d) and then increase (8-16d).

Analysis shows that returning straw to the field can adjust soil pH, reduce the damage of soil alkalinity caused by fertilization, and maintain the soil in a suitable pH environment to maintain the availability of soil nitrogen. Different types of straw returned to the field had different effects, but the difference was not significant. The decrease in soil pH during the experiment was caused by the nitrification of ammonium and the release of H^+ [19]. The soil pH decreased slowly after adding straws. On the one hand, the soil redox potential was reduced due to the flooding culture, and the high-valent iron in the soil was greatly reduced and consumed a large amount of H^+ [20]. On the other hand, it was due to the degradation of organic nitrogen. Chemicals cause the pH of the soil to rise.

Table 1. Changes of pH in soils with different treatments with time

Sample processing	0d	4d	8d	16d
-F-S	7.90a	7.66a	7.17a	7.63a
-F + M	7.20ab	6.72b	6.69b	6.84a
-F + R	6.80ab	6.89b	6.72b	6.99a
+ F-S	7.47ab	7.17c	7.10a	7.07a
+ F + M	7.21ab	7.17c	6.73b	7.31a
+ F + R	6.64b	6.86b	6.55b	7.10a

3.3 Changes Effect of Returning Straw to Field on Soil Electrical Conductivity

The soil solution has conductivity, and the strength of the conductivity can be expressed by the conductivity. Soil electrical conductivity is an indicator for determining soil water-soluble salt, and soil water-soluble salt is an important attribute of soil, and it is a factor to determine whether salt ions in soil restrict crop growth.

In this experiment, the change of soil conductivity with incubation time in different treatments is shown in Fig. 2. The order of conductivity is + F + R > + F + M > -F + R > + FS > -F + M > CK. It can be seen that returning straw to the field can increase the electrical conductivity of the soil, and the application of fertilizers has a significant impact on the electrical conductivity of the soil. With the continuation of the incubation time, CK and + FS did not change significantly, the soil conductivity of -F + M showed an upward trend, and the conductivity of + F + M and + F + R remained stable at 0-4d, and showed an upward trend at 4-8d, 8–16 days showed a downward trend.

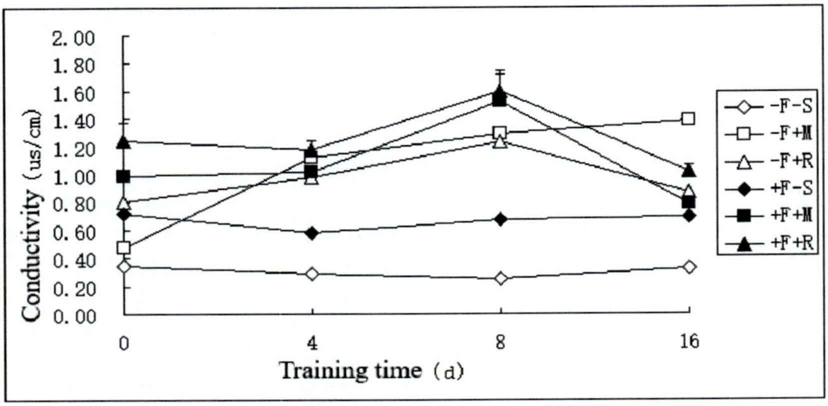

Fig. 2. Change of conductivity of soil solution with different treatments with incubation time

3.4 Changes of Soil Organic Matter After Returning Straw to the Field and Fertilizing

Soil organic matter is an important indicator to measure the level of soil fertility. The fertility of southern paddy soil is closely related to the organic matter content. A large number of studies have shown that returning straw to the field can increase soil organic matter and fertilize the soil [21–23].

The results of this experiment are shown in Fig. 3. Returning straw to the field can significantly increase the soil organic matter content. The organic matter content of the soil treated with straw increased from 0.9% before the experiment to more than 3%. The organic carbon content of paddy soil in the field treatment increased by more than 3 times on average. The order of soil organic matter on day 0 is: -F + R > -F + M > + F + M > + F + R > + F-S > CK. With the increase of incubation time, the soil organic matter content in the -F + M, + F + R and -F + R 3 treatments decreased first and then stabilized and increased slightly at the end of the incubation.

The soil ignition loss rate is an important index to measure the high temperature volatile components of the soil. In existing research, it is often used to characterize the soil organic matter content. In this experiment, the effect of adding straw on soil ash and volatile matter was considered, and the ignition loss rate was determined at the same time. On day 0, the order of soil loss on ignition is: -F + R > -F + M > + F + R > + F + M > + F-S = CK. With the continuation of the cultivation time, the burning loss rate of -F + M and -F + R soil showed the same changing trend as the corresponding total nitrogen. It shows that in the early stage of cultivation, the organic matter content of the soil decreases, which may be due to the accelerated decomposition of nitrogen-containing organic matter in the soil under flooding conditions, and a large amount of ammonia gas is released. Bernd et al. [24] showed that fresh organic materials contain a large amount of water-soluble substances, but there are big differences between different species. These substances are mainly amino acids, amino sugars, monosaccharides, polysaccharides and proteins, etc., all of which are extremely bioavailable low molecular weight compounds. Due to the limitation of the experimental environment, only harvested rice and corn stalks were used as organic materials in this experiment, but there are some differences between the two. This experiment shows that the organic matter content is very high at the initial stage of decomposing, and the decrease of organic matter after 8 days is caused by the mass reproduction and consumption of microorganisms, which may not decompose quickly due to the low temperature. But it can also be seen that there is a downward trend. Changes of soil organic matter content are shown in Fig. 3a and loss on ignition rate with culture time tare shown in Fig. 3b after different treatments. We can see that the organic matter content decreases faster when fertilizer is added to the soil. This is because the addition of fertilizer provides the necessary mineral nutrients for the growth of microorganisms, promotes the metabolic activities of microorganisms, and increases the decomposition rate. Therefore, straw and fertilizer Combined use can promote the increase of soil microbial metabolic activity and provide a favorable soil environment for crop growth.

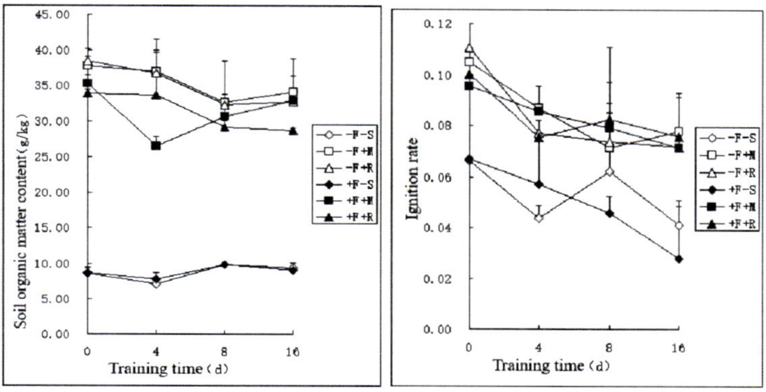

Fig. 3. Changes of soil organic matter content (Fig. 3a) and loss on ignition rate with culture time (Fig. 3b) after different treatments

3.5 Changes of Soil Total Nitrogen Content After Straw Returning to the Field and Fertilization

As shown in Fig. 4, returning straw to the field can increase the total nitrogen content of the soil, and the change of total nitrogen in the soil is different after the treatment of different types of straw returning to the field. Returning straw to the field and applying chemical fertilizers can significantly increase the total nitrogen content of the soil, and the content is in the order of -F + M > + F + M > + F + R > -F + R > + F-S > CK. Adding straw and fertilizer at the initial stage of cultivation will increase the initial total nitrogen content of the soil. The total nitrogen content of corn straw is higher than that of rice straw. During the greenhouse cultivation, the total nitrogen content of the soil showed a downward trend in all treatments. For the two treatments adding corn stalks, fertilizer had little effect on increasing the total nitrogen content of the soil after the corn stalks were returned to the field. The total nitrogen content of the soil with only corn stalks has an increasing trend in the initial stage, the total nitrogen content is the largest on the 4th day, and then there is a decreasing trend; after adding fertilizers, the total nitrogen content of the soil shows a decreasing trend during the whole process, at the end of the cultivation It is always lower than that with corn stalks only. For the two treatments of adding rice straw, the addition of fertilizer increased the total nitrogen content of the paddy soil.

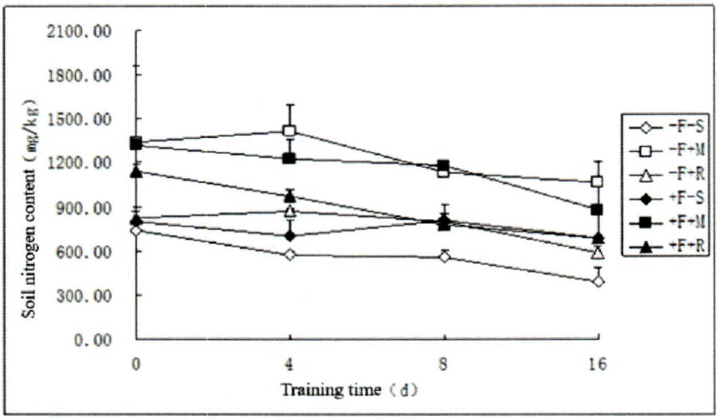

Fig. 4. Changes of soil total nitrogen content with cultivation time after different treatments

3.6 Changes of Soil Inorganic Nitrogen (Ammonium Nitrogen and Nitrate Nitrogen) After Straw Returning and Fertilization Treatments

As shown in Fig. 5, the addition of straw and fertilizer in four of the six treatments (except fertilization plus rice straw and CK) increased soil ammonium nitrogen content. The addition of fertilizers in the initial stage significantly increased the content of ammonium nitrogen. From the 4th day to the end of the cultivation period, the content of ammonium nitrogen in each treated soil tended to be stable. During the 0–16 days of culture, the group added with rice straw alone did not significantly contribute to the increase of ammonium nitrogen. The combined application of fertilization and straw increased the content of ammonium nitrogen in the soil.

As shown in Fig. 6, soil nitrate nitrogen in each group showed a decreasing trend during the whole culture process. For this experiment, adding straw treatment can effectively increase the nitrate nitrogen content of the soil. The effect of adding corn stalks alone is greater than that of adding rice stalks alone. The addition of fertilizer increased the nitrate nitrogen content in the soil treated with rice straw. For rice straw, the nitrate nitrogen content in soil decreased after adding fertilizer.

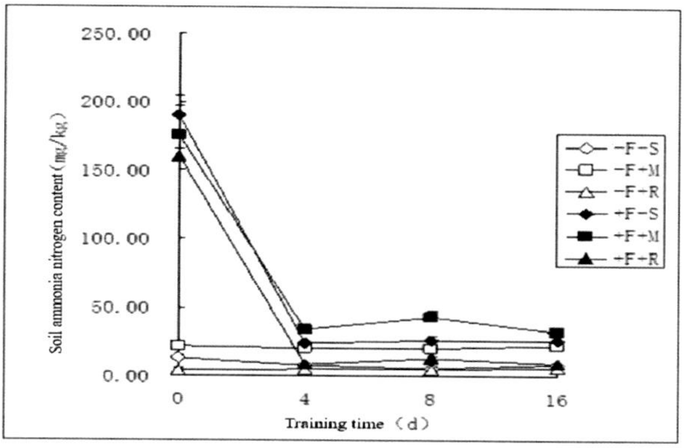

Fig. 5. Changes of soil ammonium nitrogen content in different treatments

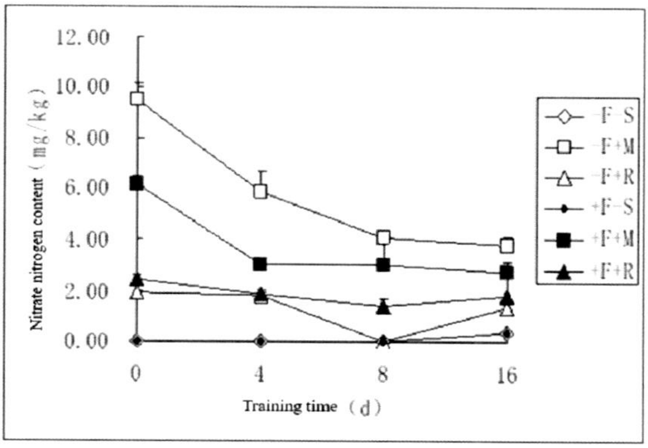

Fig. 6. Changes of soil nitrate nitrogen in different treatments

The forms of nitrogen in the soil are divided into two categories: inorganic and organic. Inorganic nitrogen is mainly ammonium nitrogen and nitrate nitrogen, which are less in soil, and nitrogen in soil mainly exists in organic form. Under normal circumstances, the inorganic nitrogen content in the soil does not exceed 5% of the total nitrogen, and this part of the nitrogen can be absorbed by crops. The intervention of a large amount of organic carbon in crop straw will significantly change the intensity and time of the soil nitrogen mineralization/immobilization process, thereby affecting the dynamic changes of inorganic nitrogen in the soil. The C/N of straw returned to the field affected this process. The contents of NH4 + -N and NO3-N in the soil with high C/N rice straw were lower than those in the soil with low C/N corn straw. Studies have shown that when a large amount of mineral nitrogen exists in the soil, the application of crop residues can improve the retention of ammonium nitrogen and nitrate nitrogen, and reduce the amount of inorganic nitrogen (especially nitrate) in the soil concentration. While using inorganic nitrogen, microorganisms also make use of nitrogen from straw, and finally decompose.

3.7 Changes of Soil C/N After Different Treatments

It can be seen from the analysis in Fig. 7 that in the early stage of cultivation, adding straw can effectively increase the C/N of the soil. The results showed that the C/N ratio of the soil showed an overall upward trend during the whole cultivation process. The changing trends of soil carbon-nitrogen ratio were different after different treatments. Under the same fertilization treatment conditions, the addition of rice straw increased the soil C/N ratio more than that of corn straw. Under the same straw treatment, the C/N ratio of each group with fertilizers was almost lower than that of groups without fertilizers. At the end of cultivation, the order of soil carbon and nitrogen ratio was: +F-S < -F-S < -F + M < +F + M < +F + R < -F + R.

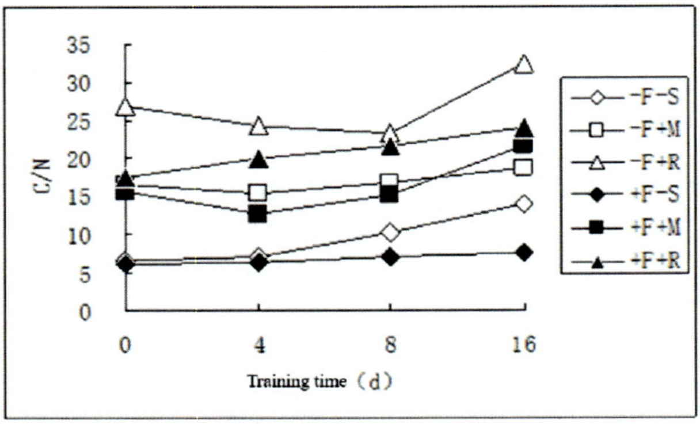

Fig. 7. Changes of C/N with culture time in different treatment

Soil carbon-nitrogen ratio is the ratio of soil carbon to nitrogen. An appropriate carbon-nitrogen ratio is helpful for microbial fermentation and decomposition. Generally speaking, the stalks of grass crops such as rice stalks, corn stalks and weeds have high carbon-nitrogen ratios. When a microorganism is decomposing organic matter, it needs to assimilate 1 part of nitrogen when it assimilates 5 parts of carbon to form its own cell body, because the carbon-nitrogen ratio of the microorganism itself is about 5:1. Therefore, it is difficult or slow for microorganisms to decompose and mineralize organic matter with a large carbon-nitrogen ratio. The optimum carbon-nitrogen ratio of microorganisms to organic matter is 25:1. If the carbon-nitrogen ratio is too large, the decomposition of microorganisms will be slow, and the available nitrogen in the soil will be consumed. Therefore, when using straw with a large carbon-nitrogen ratio (such as corn straw, rice straw, etc.) to return to the field, fertilizers containing more nitrogen should be supplemented to adjust the carbon-nitrogen ratio.

3.8 The Effect of Each Factor on the Experiment

As shown in Table 2: the influence of each factor and the interaction between each factor on each index of this experiment is significant. Among them, cultivation time had a significant effect on organic matter and loss on ignition rate, and had a very significant effect on other indicators; fertilization had no significant effect on total nitrogen content, which might be the main effect of straw addition. Because the cultivation time is short, the straw is not completely decomposed.

Table 2. The significance test of the influence of cultivation time, fertilization and straw treatment on the main basic physical and chemical properties of soil

Index	pH	Conductance	Organic matter	Total nitrogen	Ammonium nitrogen	Nitrate nitrogen	Ignition rate
Time	< 0.001**	< 0.001**	< 0.05*	< 0.001**	< 0.001**	< 0.001**	< 0.05*
Fertilize	0.052	< 0.001**	< 0.001**	0.109	< 0.001**	< 0.001**	0.719
Straw	< 0.001**	< 0.001**	< 0.001**	< 0.001**	< 0.001**	< 0.001**	< 0.05*
Time × Fertilize	0.233	< 0.001**	0.428	0.855	< 0.001**	< 0.001**	0.948
Time × Straw	< 0.001**	< 0.001**	< 0.05*	0.725	0.073	< 0.001**	0.850
Fertilize × Straw	< 0.001**	< 0.001**	< 0.05*	< 0.05*	< 0.05*	< 0.001**	0.609
Time × Fertilize × Straw	< 0.05*	< 0.001**	0.163	0.591	0.507	< 0.05*	0.644

4 Conclusion

In conclusion, straw returning to the field can increase the organic carbon content of paddy soil, improve soil fertility, help soil environment remediation, and protect soil ecology. This paper draws the following conclusions, in order to provide a scientific basis for rationally returning straw to the field and making full use of straw resources in production practice.

(1) Returning straw to the field can significantly increase the content of soil organic matter and improve soil fertility. For paddy soil, returning rice straw to the field without fertilization is better. The combined application of chemical fertilizers can effectively reduce the carbon-nitrogen ratio, promote the decomposing of the straw in the soil, facilitate the decomposition and release of the straw's own nutrients, and avoid the competition of decomposing bacteria for nitrogen. The corn straw combined fertilization treatment has the most significant effect.

(2) When using straw with a large carbon-nitrogen ratio to return to the field (such as corn straw, rice straw, etc.), an appropriate amount of fertilizer containing more nitrogen should be supplemented to adjust the soil carbon-nitrogen ratio. At the same time, attention should be paid to the sequence of fertilization and straw returning to avoid excessive nitrogen loss and secondary pollution.

(3) In order to make full use of straw returning to the field to protect the soil ecological environment, when using rice straw with a large carbon-nitrogen ratio, fertilizers containing more nitrogen should be supplemented to adjust the carbon-nitrogen ratio and make the carbon-nitrogen ratio reach an appropriate value.

References

1. Gao, J., Sun, Z., Feng, L., et al.: Effects of combined application of straw and nitrogen fertilizer on soil enzyme activities and soil nutrients in arid regions of western Liaoning. Chinese J. Ecological Environ. **21**(4), 677–681 (2012)
2. Bai, J., Li, Y., Zhang, J., et al.: Straw returning and one time application of a mixture of controlled release and solid granular urea to reduce carbon footprint of plastic film mulching spring maize. J. Clean. Prod. **280**, 124478 (2021)
3. Kuai, W., Zhang, L., Song, Y., et al.: Effects of stablenitrogen fertilizer combined with straw returning on rice yield and N2O and CH4 emissions. Chin. J. Appl. Ecol. **30**(4), 1287–1294 (2019)
4. He, T., Wang, J., Yunpeng, F.: Effects of adding straw and biochar with equal carbon content on soil respiration and microbial biomass carbon and nitrogen. Environ. Sci. **42**(1), 450–458 (2021)
5. Gao, J., Sun, Z., Feng, L., et al.: Effects of combined application of straw and nitrogen fertilizer on maize growth and water use efficiency. J. Northeast. Agric. Univ. **42**(11), 116–120 (2011)
6. Wang, R., Zhang, J., Dong, S., et al.: Current situation and effect of straw resource utilization in main maize producing areas in my country. J. Appl. Ecol. **22**(6), 1504–1510 (2011)
7. Zhang, X., Li, J., Meng, F., et al.: Ecological benefit analysis of long-term return of winter wheat and summer maize straws to the field in Huantai County. Chin. J. Ecol. **40**(1), 4157–4168 (2020)
8. Huang, D., Zhu, P., Wang, Z., et al.: Discussion and doubt on the decomposition rate of organic carbon in dry and paddy fields. Soil J. **35**(4), 482–492 (1998)
9. Tomohiro, B., Clive, B.L.: Soil solid and solution phase change and mung be an response during amelioration of aluminium toxicity with organic matter. Plant Soil **140**(2), 183–196 (1992)
10. Chen, X., Zhou, J.: Effects of different fertilization treatments on carbon-nitrogen ratio and mineralization characteristics of corn stover. Chin. J. Appl. Ecol. **20**(2), 314–319 (2009)
11. Ding, X., He, H.: Effects of different nitrogen supply levels on carbon mineralization and microbial biomass at the initial stage of corn stover degradation. Soil J. **39**(4), 35–39 (2008)
12. Chen, S., Liu, D., Zhu, Z., et al.: Effects of straw returning and no-tillage on soil nutrients and carbon pools in Sichuan Basin. China Soil and Water Conservation Science6 (Supplement), 54–56 (2008)
13. Hong, C.: Study on the impact of direct return of full straw to soil fertility and farmland ecological environment. J. Zhejiang Univ. (Agric. Life Sci. Ed.) **29**(6), 627–633 (2003)
14. Zhijie, W., Zhang, H., Guangshan, X., et al.: Effect of returning corn stalks to field for fertilizing soil. Chin. J. Appl. Ecol. **13**(5), 539–542 (2002)
15. Wang, S.: Effects of adding corn residue on nitrogen transformation in soil-plant system. Chin. J. Appl. Ecol. **15**(3), 449–452 (2004)
16. Li, X., Ji, W., Zhu, H., et al.: Effects of straw returning on crop yield and soil nutrients. Anhui Agric. Sci. **31**(5), 870–871 (2003)
17. Glendining, M.J., Powlson, D.S., Poulton, P.R., et al.: The effects of long-term applications of inorganic nitrogen fertilizer on soil nitrogen in the Broadbalk Wheat Experiment. J. Agric. Sci. **127**, 347–363 (1996)

18. Jin, Y., Sun, Y., Ke, F., et al.: Mechanism of rice yield increase by returning organic nitrogen from wheat straw. Shanghai Agric. J. **7**(2), 62–66 (1991)
19. Feng, Y., Sven, S., Konrad, M.: Soil pH increase due to biological decarboxylation of organicanions. Soil Biol. Biochem. **28**, 617–624 (1996)
20. Hue, N.V., Amien, I.: Aluminum detoxification with green manures. Commun. Soil Sci. Plant Anal. **20**, 1499–1511 (1989)
21. Hou, Z., Guijuan, D., Sun, Z., et al.: Research on the effect of returning corn stalks to the field for fertilization. Miscellaneous Grain Crops **24**(3), 166–167 (2004)
22. Whitbread, A., Blair, G., Konboon, Y., et al.: Managing crop residues, fertizers and leaf litters to improve soil C, nutrient balances, and the grain yield of rice and wheat cropping systems in Thailand and Australia. Agr. Ecosyst. Environ. **100**, 251–263 (2003)
23. Yadvinder-Singh, B.-S., Ladha, J.K., Khind, C.S., et al.: Effects of residue decomposition on productivity and soil fertility in rice-wheat rotation. Soil Sci. Soc. Am. J. **68**(3), 854–864 (2004)
24. Marscher, B., Noble, A.D.: Chemical and biological processes leading to the neutralisation of acidity in soil incubated with litter materials. Soil Biol. Biochem. **32**, 805–813 (2000)

Expression and Characterization of Dichloromethane Dehalogenase for Dichloromethane Degradation

Yicheng Wu(✉), Chen Yu, Haiyan Fu, and Huichao Yao

College of Environmental Science and Engineering, Xiamen University of Technology, Xiamen, China

ycwu@xmut.edu.cn

Abstract. Dehalogenases widely used to remediate halocarbons contaminated soil. Genetic modification and over expression of recombinant dehalogenases will greatly facilitate its application in environment remediation area. In this study, a gene encoding dehalogenase (LinB) was modificated and over expressed in Pichia pastoris expression system. The results showed that the intracellular recombinant enzymatic activity were most active when the recombinant bacterium cultured at 30 °C for 4 h with glycerin 10 g/L as inducer, and the removal rate of 2-choloroethy ethyl sulphide by intracellular enzyme solution was 86.44% .

Keywords: Dichloromethane dehalogenase · Dichloromethane · *Pichia pastoris* · Enzymatic activity

1 Introduction

The widespread use of halogenated organic compounds as pesticides, herbicides and other industrial products has led to their wide distribution in the environment [1–4]. Because of extraordinary stability of the chemical bonds, the variety of halogenated substances and their derivatives are found in soils, groundwater aquifers and sediments with the greatest frequency and the highest concentration throughout the world. Occurrence of halogenated substances in the environment is of great concern due to their hazardous nature and persistent environmental pollution [5].

Dehalogenase plays an important role in the remediation of halogenated hydrocarbons contaminated water with the advantages of cost-effectiveness and ecological friendliness. Dehalogenase produced microorganism metabolism has broad application prospects in the field of halogenated hydrocarbons contaminated water remediation [6–8]. Isolation of new microorganisms with high enzyme production and activity is of great significance for the application of dehalogenase in the field of environmental remediation. However, the removal performance of chlorinated hydrocarbons by natural microorganisms is very low due to their high toxicity and hard degradation. The content of halogenated hydrocarbons in the environment is increasing year by year. Overexposure to such halogenated hydrocarbons causes toxic and lethal accumulation leading to severe environmental and health consequences [9].

X. Chen (Ed.): ICESE 2022, ESE, pp. 50–56, 2023.

The removal performance of chlorinated hydrocarbons by natural microorganisms is very low due to their high toxicity and hard degradation.

The growing interest in the use of microbial processes in the removal and recovery of halogenated toxic polluted environments is practically due to their inducible enzyme system producing dehalogenases. More and more attentions have pained to environmentally relevant enzymes involved in biodegradation of these pollutants. Catalytic efficiency determines the cost and effect of chloromethane contaminated sites remediation. The application of protein engineering tools in mutagenesis studies and sequence analysis improved or modified dehalogenase activities. [10, 11]. In the present study, over expression of dehalogenases in *Pichia pastoris* and 2-chloroethyl ethyl sulfide (2-CEES) removal activity were investigated.

2 Methods and Materials

2.1 Strains and Reagents

Pichia pastoris strain GS115 and pPICZα were provided by Fujian Normal University, pMD-19 was purchased from Takara Biotech (Dalian, China). DNA marker and polymerase chain reaction primers were purchased from Sangon Biotech (Shanghai, China). 2-CEES bought from Aladdin Industrial Corporation.

2.2 Over Expression of Dehalogenase Gene in *Pichia Pastoris*

According to dehalogenase gene sequence from patent US7888103B2, we have synthesized the dehalogenase gene with restriction enzyme sites of *Xba* I and Xho I. The synthesized gene was subsequently ligated to pMD-19 constructed pMD-19-linB vector and transformed into *E. coli* DH5a. After pMD-19-linB *Xba* I and Xho I digestion, the released linB gene was subsequently cloned into *Xba* I-Xho I restriction site of expression vector pPICZα to generate pPICZα-linB vector. The inserted gene sequence was confirmed by *Xba* I and *Xho* I digestion together with direct DNA sequencing. The recombinant plasmid pPICZα-linB was transformed into *Pichia pastoris* for expression through electrotransformation with methanol as an inducer.

2.3 Crude Dehalogenase Extraction

The precipitated cells were collected from the centrifugation, and the cell in the centrifuge tube was washed with 0.1 M phosphate buffer at pH 7.0. Re-suspended cells are repeatedly frozen and thawed, and then broken by ultrasonic wave with a sonicator. Broken cell solution centrifuged at 10,000 rpm for 3 min to obtain the supernatant as intracellular crude dehalogenase.

2.4 Analysis of Dehalogenase Activity

Analysis of dehalogenase activity was performed using an gas chromatograph (GC-2010, Shimadzu, Japan) system on a HP-5 column with a flame ion detector. The inlet is 250 °C for gasification. The column heating program is 40 °C for 3 min, then rose to 200 °C at a rate of 15 °C/min, and kept for 1 min.

2.5 2-CEES Removal Effect Evaluation

Add 5 mL of culture supernatant into 5 colorimetric tubes, and then 5 mg of 2-CEES was added respectively. Dichloromethane was added to stop the reaction to control the reaction time for 10 min. Take the lower layer of supernatant and filter before GC analysis. Use YPD culture medium as the control, and calculate the removal rate of culture supernatant at different induction times. The calculation formula of removal rate (r) is as follows:

$$r = \frac{C_0 - C_t}{C_0} \times 100\%$$

where, C_0 and C_t are the initial mass of 2-CEES and the mass after treatment t time, respectively.

3 Results and Discussion

3.1 Standard Curve Drawing

Different concentrations of 2-chloroethyl ethyl sulfide (2-CEES) were determined by gas chromatography. Taking the 2-CEES concentration as the ordinate and the peak area as the abscissa to draw the standard curve (Fig. 1). The relationship between 2-CEES

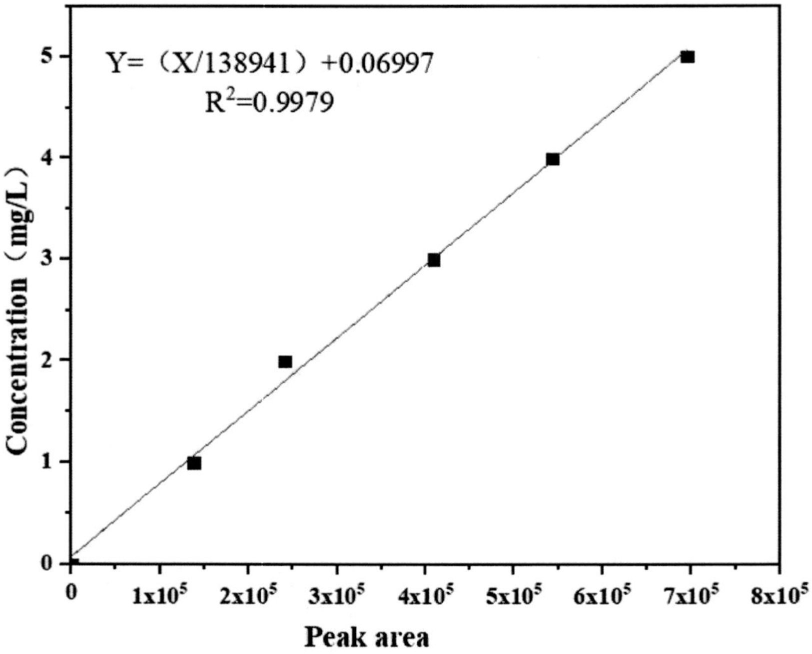

Fig. 1. Standard curve of 2-CEES

concentration(Y, g/L) and peak area (X) is shown as Y = X/138941 + 0.06997, and R^2 is 0.9979, indicating that the peak area has a good linear relationship with 2-CEES concentration.

3.2 Detection of Dehalogenase Activity in Recombinant Strain

Recombinant strain *Pichia pastoris* with pPICZα-linB was picked out for induction, the wide type cells were collected as controls. Dehalogenase activity in recombinant strain was shown in Fig. 2. As shown in Fig. 2, the removal rate of 2-CEES in the control wild-type cell disruption solution was only 10.01%, and the highest removal rate of recombinant cell disruption reached to 76.98%. However, the recombinant culture supernatant has a certain 2-CEES degradation (27.08%), which may be related to the release of dehalogenase by partial recombinant cell lysis. All these results indicated that dehalogenase functionally expressed in *Pichia pastoris.*

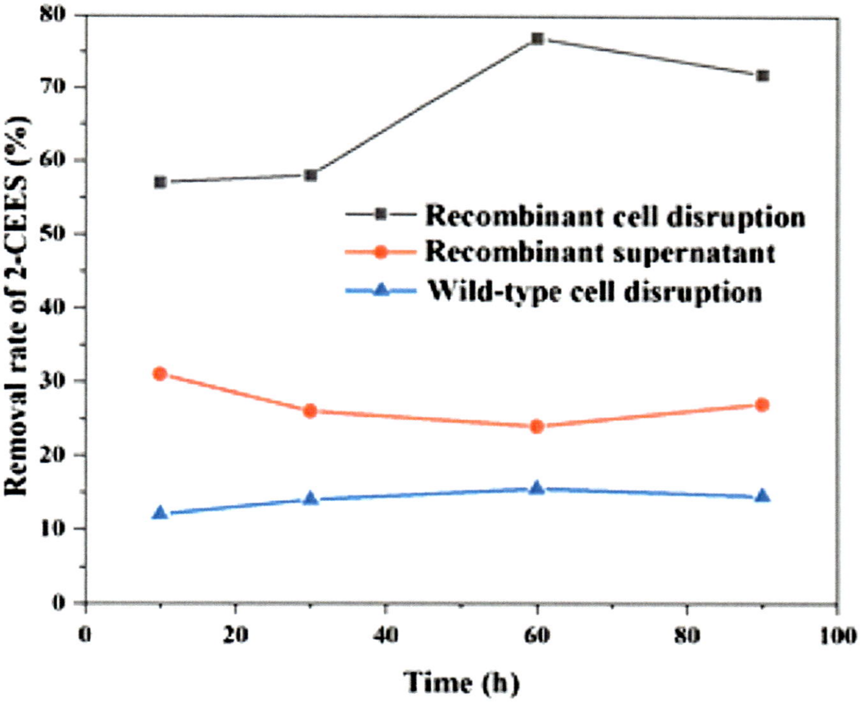

Fig. 2. Removal efficiency of 2-CEES by recombinant strain

3.3 Effect of Glycerol Concentration and Induction Time on Intracellular Dehalogenase Activity

Different concentrations of glycerol were added as an inducer for *dehalogenase* expression, when the recombinant suspension was grown to absorbance reached to 1.2 at a wavelength of 600 nm. The effect of glycerol concentration and induction time on intracellular enzyme activity was induced at 30 °C was shown in Fig. 3. As shown in Fig. 3, the 2-CEES removal rate increases with the increase of glycerol concentration after induction for 2 h, and the highest removal rate reached to 70.39%. The 2-CEES removal rate of 4 h induction is significantly higher than other induction time. The removal efficiency of 2-CEES was highest at glycerol concentrations of 10 g/L induced for 4 h, and the removal rates of 2-CEES was 86.44%.

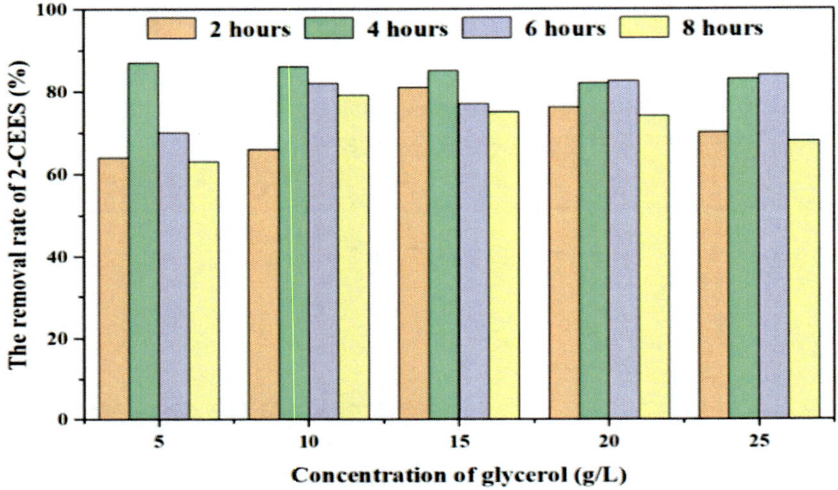

Fig. 3. Effect of glycerol concentration and induction time on 2-CEES removal efficiency

3.4 Effect of Inducer Concentration and Induction Time on Intracellular Protein Content of Recombinant Strain

The effects of glycerol concentration and induction time on intracellular proteins of recombinant strain are shown in Fig. 4. As shown in Fig. 4, the lowest intracellular protein content was induced for 2 h with 10 mg/L glycerol as inducer. Intracellular protein content in the recombinant bacteria was the highest when the glycerol induced for 4 h at concentration of 10 g/L, and the intracellular protein concentration reached to 620.51 μg/mL. The intracellular dehalogenase activity was also the highest under such induction condition, and the protein content was consistent with the enzyme activity.

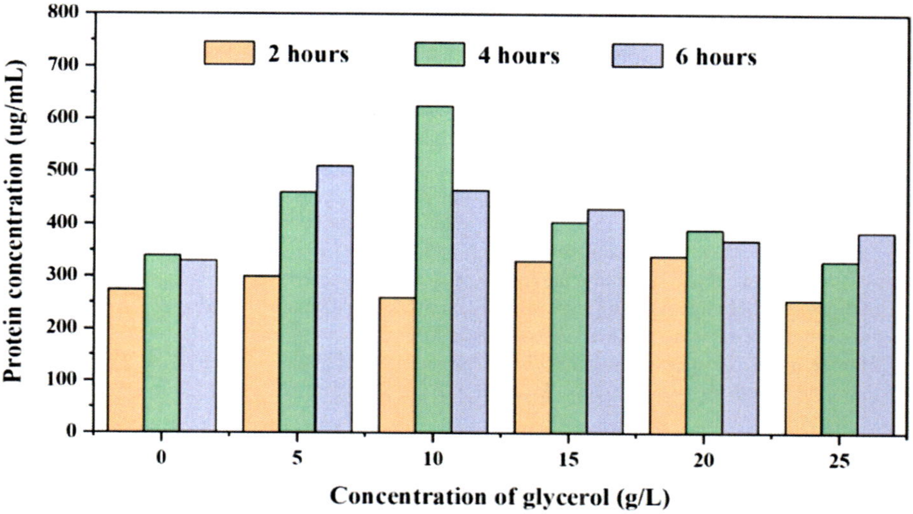

Fig. 4. Effect of glycerol on protein content in recombinant strain

4 Conclusions

This study successfully ligated the dehalogenase coding sequence into vector pPICZα, and transformed into *Pichia pastori* to construct expression recombinant strain. The expressed enzyme activity assay showed that 2-CEES removal rate reached to 86.44% by the broken cells of *Pichia pastori* within pPICZα-linB. The highest enzyme activity and intracellular protein content were obtained after induction at 30 °C for 4 h when glycerol with a final concentration of 10 g/L.

Acknowledgements. This work was supported by Ximen Natural Science Foundation of (3502Z20227214).

References

1. Negri, A., Marco, E., Damborsky, J., Gago, F.: Stepwise dissection and visualization of the catalytic mechanism of haloalkane dehalogenase LinB using molecular dynamics simulations and computer graphics. Mol. Graph. Model **26**, 643–651 (2007)
2. Chaudhry, R.G., Chapalamadugu, S.: Biodegradation of halogenated organic compounds. Microbiol. Rev. **55**, 59–79 (1991)
3. Belkin, S.: Biodegradation of haloalkanes. Biodegradation **3**, 299–313 (1992)
4. Bidmanova, S., Chaloupkova, R., Damborsky, J., Prokop, Z.: Development of an enzymatic fibre-opticbiosensor for detection of halogenated hydrocarbons. Anal. Bioanal. Chem. **398**, 1891–1898 (2010)
5. Vatsal, A.A., Zinjarde, S.S., Ameeta, R.K.: Phenol is the initial product formed during growth and degradation of bromobenzene by tropical marine yeast, yarrowia lipolytica NCIM 3589 via an early dehalogenation step. Front. Microbiol. **8**, 1165 (2017)

6. Copley, S.D.: Diverse mechanistic approaches to difficult chemical transformations: microbial dehalogenation of chlorinated aromatic compounds. Chem. Biol. **4**, 169–174 (1997)
7. De Jong, R.M., Dijkstra, B.W.: Structure and mechanism of bacterial dehalogenases: different ways to cleave a carbon-halogen bond. Curr. Opin. Struct. Biol. **13**, 722–730 (2003)
8. Kurihara, T., Esaki, N.: Bacterial hydrolytic dehalogenases and related enzymes: occurrences, reaction mechanisms, and applications. Chem. Rec **8**, 67–74 (2008)
9. Puzyn, T., Haranczyk, M., Suzuki, N., Sakurai, T.: Estimating persistence of brominated and chlorinated organic pollutants in the air, water, soil, and sediments with the QSPR-based classification scheme. Mol. Divers. **15**, 173–188 (2011)
10. Raina, V., et al.: Hydroxylated metabolites of β- and δ-hexachlorocyclohexane: bacterial formation, stereochemical configuration, and occurrence in groundwater at a former production site. Environ. Sci. Technol. **41**, 4291–4298 (2007)
11. Manickam, N., Mau, M.,Schlömann, M.: Characterization of the novel HCH-degrading strain, microbacterium sp. ITRC1. Appl. Microbiol. Biotechnol. **69**, 580–588 (2006)

Resource Management and Ecological Environment Protection

Optimal Solution for Increasing Groundwater Pumping by Integrating MODFLOW-USG and Particle Swarm Optimization Algorithm: A Case Study of Wadi El-Natrun, Egypt

Ahmed M. Saqr[1,2](✉), Mahmoud Nasr[1,3], Manabu Fujii[4], Chihiro Yoshimura[4], and Mona G. Ibrahim[5]

[1] Environmental Engineering Department, Egypt-Japan University of Science and Technology, E-JUST, Alexandria, Egypt
ahmed.saqr@ejust.edu.eg
[2] Irrigation and Hydraulics Department, Faculty of Engineering, Mansoura University, Mansoura, Egypt
ahmedsaqr@mans.edu.eg
[3] Sanitary Engineering Department, Alexandria University, Alexandria, Egypt
[4] Department of Civil and Environmental Engineering, Tokyo Institute of Technology, Tokyo, Japan
[5] Environmental Health Department, High Institute of Public Health, Alexandria University, Alexandria, Egypt

Abstract. Extensive use of groundwater causes over-exploitation of aquifers globally. Hence, this study aims at introducing a simulation-optimization (S-O) framework of the MODFLOW-USG model and the particle swarm optimization (PSO) algorithm for groundwater sustainability in water-stressed regions. This S-O framework was applied in the Wadi El-Natrun basin, Egypt, which suffers from a continuous drop in groundwater heads. After successful calibration, the simulation model was used to predict the changes in groundwater heads due to the increase in the existing pumping rates by 25% (Scenario No. 1) and 50% (Scenario No. 2) till 2030. Then, the PSO algorithm was integrated with the MODFLOW-USG model through a MATLAB environment to get the optimum discharges under specified drawdown values in groundwater heads. The results of the simulation model exhibited that the existing total extraction rate from the study domain was equal to 227000 m^3/day (Base value). By 2030, increasing the existing pumping rates due to the two scenarios might have a substantial effect on lowering the groundwater levels in the study area, especially the southwestern portions. Based on the S-O results, the optimum overall pumping rates were expected to be 266556 m^3/day (117% of the base value) and 307413 m^3/day (135% of the base value) to assign drawdowns in groundwater heads to be less than or equal to the median values at all wells, respectively. The proposed S-O methodology could be useful for the optimal exploitation of groundwater in the study area and other comparable basins worldwide to partially meet sustainable development goals (SDGs).

Keywords: Simulation-optimization (S-O) · MODFLOW-USG · Particle swarm optimization (PSO) · Groundwater head drawdowns · Sustainable development goals (SDGs)

X. Chen (Ed.): ICESE 2022, ESE, pp. 59–73, 2023.

1 Introduction

Groundwater is considered one of the most precious water resources in arid and semi-arid regions. Moreover, it can be used for various purposes, including domestic, agricultural, and industrial needs [1, 2]. However, the world faces a water crisis and many aquifers have witnessed a continuous decline in groundwater heads owing to unplanned horizontal expansions [3]. Therefore, there is an urgent need for efficient management strategies to preserve groundwater at an optimal state [4]. Many researchers have used several techniques for the sustainable use and supply of groundwater. Simulation-optimization (S-O) framework can offer influential solutions to the harmful effects of increasing groundwater extraction rates. This technique is formed by coupling a simulation model with an optimization algorithm to benefit from the capabilities of these two components [5]. The development and implementation of simulation models and optimization algorithms for the management of groundwater systems have become fruitful research subjects in recent decades [6].

On the one hand, different numerical models were applied to simulate and predict the physical behavior of the groundwater flow regimes. However, MODFLOW was the most common model among studies. This model was first developed by the United States Geological Survey (USGS) in the early 1980s [7]. It uses a rectangular finite-difference cell-centered grid to facilitate the three-dimensional simulations of groundwater. Unfortunately, it exhibits some drawbacks, including domain discretization inaccuracy and approximate well locations throughout the grid cells [8]. A recent unstructured grid version of the MODFLOW model, called MODFLOW-USG, was released in 2013. The MODFLOW-USG model supports a wide variety of non-rectangular grid types and uses a generalized control finite volume technique to model groundwater flow. So, unlike prior MODFLOW versions, it allows for the creation of irregular model fields and better refinement around wells at their original locations. Consequently, it can lead to a high level of accuracy and quicker solutions [9].

On the other hand, optimization algorithms can be applied to get optimal pumping rates for different management scenarios that cannot be obtained using numerical models. The metaheuristic type has recently become popular because of its ability to solve complex nonlinear groundwater problems and its flexible linking with numerical models [6]. The particle swarm optimization (PSO) algorithm is considered an intelligent metaheuristic algorithm and was first introduced by Eberhart and Kennedy (1995) [10]. It is characterized by simplicity, flexibility, and a derivation-free mechanism [11]. Many researchers have assessed the performance of the PSO algorithm over other metaheuristic algorithms. For instance, Matott et al. (2006) [12] compared the results of the genetic algorithm (GA), conjugate gradient (CG), simulated annealing (SA), and PSO to get the optimal discharges of wells for two different research studies. It was concluded that the PSO algorithm led to the best outcomes with less computational time. Also, comparable results were attained by many investigators using the PSO algorithm to solve groundwater optimization problems, showing its high accuracy and the avoidance of local optimum solutions [13].

The S-O framework, formed of the MODFLOW-USG model and the PSO algorithm, has been successfully adopted in recent studies for solving the calibration problem of numerical models [7]. However, its application to get the best groundwater withdrawals from aquifers under extensive use has not been studied yet. Hence, this research study aimed at developing this S-O technique for optimal and sustainable exploitation of groundwater in water-stressed areas. The developed S-O framework was applied in the Wadi El-Natrun basin, Egypt. This basin was selected to carry out the current research because it has witnessed a deterioration in groundwater levels owing to the increase in groundwater extraction that may take place in other similar regions all over the world. The changes in groundwater heads were predicted using the MODFLOW-USG model for two scenarios of increasing the existing pumping rates by 25% and 50% till 2030. These scenarios were anticipated to fulfill the future water needs for additional reclamation expansions. Then, the PSO algorithm was linked to the simulation model through a MATLAB environment to get the optimum extraction rates under the constraints of median drawdown values in groundwater heads obtained from the forecasting scenarios. Next, the findings of the S-O framework were linked to many environmental, economic, and social benefits associated with sustainable development goals (SDGs). The proposed objectives can help decision-makers determine optimal groundwater pumping strategies to meet extra demands in the future.

2 Materials and Methods

2.1 Study Area Description

The Wadi El-Natrun basin, in Egypt's Behera governorate, is situated on the southwestern outskirts of the Nile Delta. It lies between longitudes: (30°01′05″E - 30°30′35″E) and latitudes: (30°16′24″N–30°31′59″N), as shown in Fig. 1. It has a length of nearly 55 km and a width of around 12 km along the Cairo-Alex Desert Road to cover an area of about 485.25 km^2. It is considered a semi-arid region with dry hot summers and mild rainy winters. Eight drainage lakes occupy the floor of the basin [4]. The Pliocene aquifer within Wadi El-Natrun provides people there with most of their water needs. Inferred faulting systems bound it from all directions except for the northern direction, representing its extensions. It mainly gets its water from the neighboring Pleistocene aquifer via the northeastern inferred fault line. Besides, the general flow is from the northeast to the southwest [14]. Also, there are other contributions from local rainstorms and infiltration from excess irrigation [15].

2.2 The Framework of the Stepwise Methodology

The following methodology steps were applied:

Data Collection. Different types of available data were collected to construct a database for the study area. The details of this data with its sources were:

- The digital elevation model (DEM) of the basin under study [16].
- Location of faults line [14] and pumping wells taping the Pliocene aquifer [15, 17, 18], as demonstrated in Fig. 1.
- A hydrogeological cross-section showing the decomposition of the underlying successive soils, as illustrated in Fig. 2(a) [19].

- Measured groundwater heads in January 2015 [15] and in December 2020 from a field survey based on 34 observation wells, as depicted in Fig. 2(b).
- Hydraulic conductivity and specific storage in the range of (0.40–41.34) m/day, and (0.0001–0.1), respectively [14, 20].
- The maximum effective surface recharge of approximately 55.1 mm/year [18].
- The mean evaporation rate from the drainage lakes of about 630 mm/year [21].

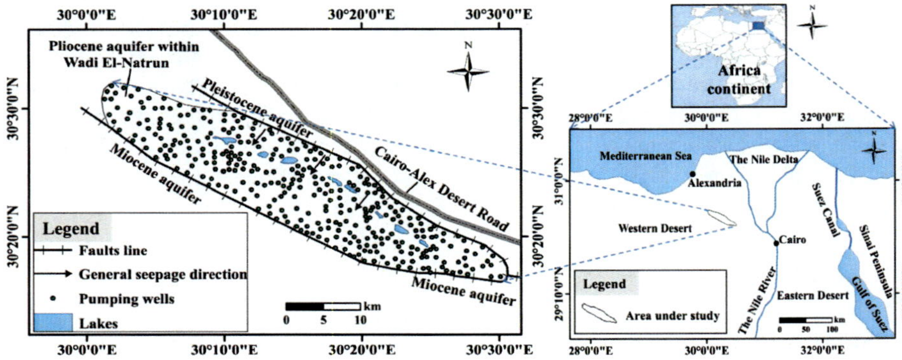

Fig. 1. Location map of the basin under study.

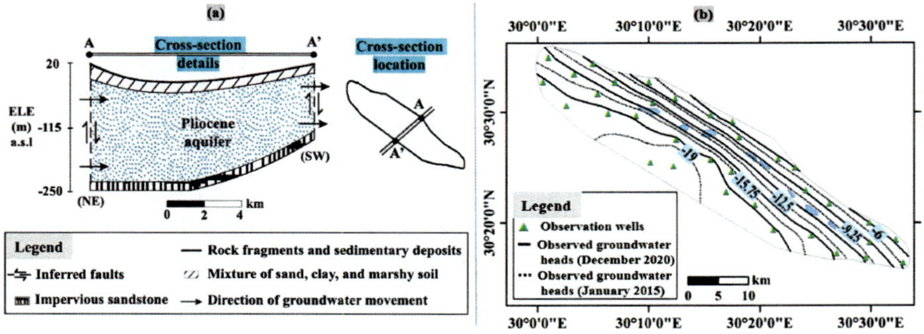

Fig. 2. Available data for the basin under study: (a) a hydrogeological cross-section, and (b) observed groundwater heads.

Data Preprocessing. Groundwater modeling system (GMS) is a powerful graphical user interface that can be used for different purposes of groundwater simulations [9]. Thus, it was employed to construct the conceptual model of the Wadi El-Natrun basin based on in-built geographic information system (GIS) tools. First, the georeferenced image in Fig. 1 was inserted into GMS to act as a base map to delineate different elements in the study domain. Second, the boundary conditions of the model were conceptualized. The lateral boundaries of the basin and interior lakes were assigned as general head boundaries. This type of assignment could permit groundwater to flow into and out of the model field at a rate that might be proportional to the difference between groundwater heads. Third, top surface elevations were extracted from the DEM. Then, the details of the hydrogeological cross-section were used to define the vertical extensions of the three geological layers. Next, different integrated packages in GMS were utilized to define sources and sinks. Sources represented recharge amounts, while sinks included pumping and evaporation rates. After that, coverages were created for the aquifer properties, including hydraulic conductivity and specific storage. Finally, a 3D unstructured Voronoi grid was generated for the domain under study, as shown in Fig. 3. This type of grid provides more elasticity to the design of model cells [8].

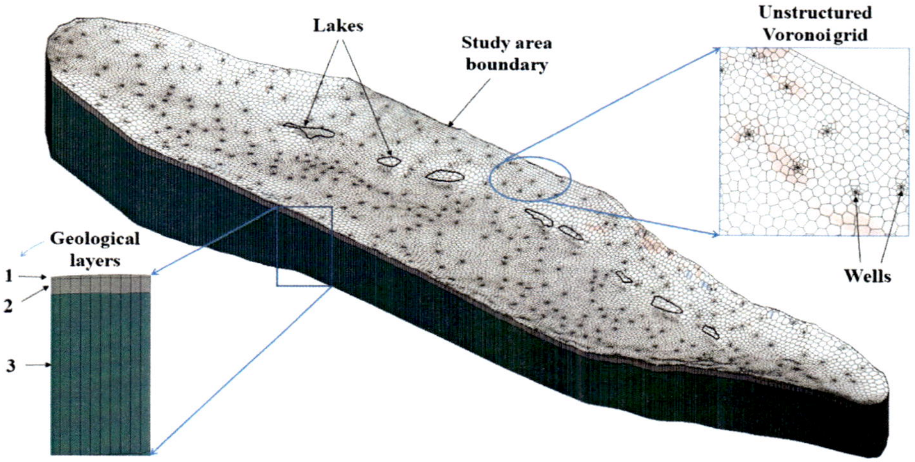

Fig. 3. A 3D view of the conceptual model of the basin under study.

Following the construction of the conceptual model, it was exported to a numerical MODFLOW-USG model through GMS. The following partial differential equation was applied by the MODFLOW-USG model to mathematically simulate the groundwater behavior [11]:

$$\frac{\partial}{\partial x}\left(K_{xx}\frac{\partial h}{\partial x}\right)+\frac{\partial}{\partial y}\left(K_{yy}\frac{\partial h}{\partial y}\right)+\frac{\partial}{\partial z}\left(K_{zz}\frac{\partial h}{\partial z}\right)\pm w=S_s\frac{\partial h}{\partial t} \tag{1}$$

where: K_{xx} & K_{yy} & K_{zz} are the hydraulic conductivities in the x & y & z directions, respectively, h is the hydraulic head, w represents the sources and/or the sinks, S_s is the specific storage, and t is the time.

Calibration of the Simulation Model. After preparing the simulation process, the MODFLOW-USG model should be calibrated to be used for any application. Model calibration is the procedure of systematically altering some parameters and repeatedly running the model until the estimated output matches the data observed in the field within an acceptable level of accuracy [15]. To calibrate the Wadi El-Natrun simulation model between January 2015 and December 2020, a trial-and-error process was applied in two steps:

- The MODFLOW-USG model was first calibrated in the steady phase. Hydraulic conductivity, along with different recharge and discharge parameters, were adjusted within possible ranges to decrease the difference between the calculated and measured groundwater heads in January 2015 at the observation wells.
- Then, it was calibrated in the transient stage. The initial conditions were the groundwater heads obtained from the steady-state model. Then, the storage coefficient, in addition to various recharge and discharge factors, were modified to get simulated groundwater heads by December 2020 very close to the observed ones.

Validation of the Simulation Model. At the end of the transient calibration, the 'Flow Budget' tool within GMS was applied to get the existing total pumping rate from the Pliocene aquifer of the study area. The adjusted simulation model was then used to predict changes in groundwater heads due to increasing the existing extractions by 25% (Scenario No. 1) and 50% (Scenario No. 2) till December 2030. The spatial distributions of the predicted groundwater heads were mapped with the 'UGrid' module for the two scenarios. The values of predicted groundwater heads were compared with those of 2020 to compute the forecasted extra drawdown values at the observation wells. Box plots were drawn using Microsoft Excel software to display different statistics about the drawdowns, including minimum, median, and maximum values.

Formulation of the S-O Framework. A proposed S-O framework was formed in a MATLAB environment to get the optimal pumping rates from the study area for two different cases. Median values of drawdown, obtained from the two management scenarios, were assigned as constraints to protect the ecosystem from further deterioration. The S-O framework's structure could be illustrated in the following lines (Fig. 4):

- The S-O framework firstly started the search for the optimal solution by generating a sample of random particles, i.e., solutions. Moreover, each particle in the search space was identified by its position and velocity. Accordingly, each particle flies towards the best particle at a rate determined by its current position. This procedure physically resembles a swarm of birds communicating as they fly. Each bird looks in a specific and then communicates with the others to determine which bird is in the best spot direction [22].
- The S-O framework was then attempted to calculate the value of the fitness function, which aimed at maximizing the overall pumping (Q_t) from the study area as:

$$Q_t = \max\left\{\sum_{i=1}^{N_w} Q_i - P\right\} \tag{2}$$

where: N_w is the total number of pumping wells, Q_i is the quantity of pumping from each well which is bounded between Q_i^{min} (Minimum value of pumping) and Q_i^{max} (Maximum value of pumping), and P is a penalty term related to the drawdown constraint.

The estimation process of the fitness function was performed by running the MODFLOW-USG model as a subroutine in the S-O framework subjected to the following drawdown constraint:

$$F_i \le D_i \tag{3}$$

where: F_i is the computed drawdown at each well, and D_i is a specified value of drawdown (Median value computed from each management scenario).

It should be noted that P was equal to zero if the constraint was not violated. Otherwise, a penalty value was considered and added to the fitness function. After that, the final value of the fitness function was displayed, and this stage of the S-O framework was terminated.

- In this step, the position and velocity of each particle were updated for each iteration in the following way [23]:

$$v_i^{ite+1} = r_1 v_i^{ite} + c_1 r_2 \left(x_{best_i^{ite}} - x_i^{ite} \right) + c_2 r_3 \left(x_{gbest_i^{ite}} - x_i^{ite} \right) \tag{4}$$

$$x_i^{ite+1} = x_i^{ite} + v_i^{ite+1} \tag{5}$$

where: r_1 & r_2 & r_3 are uniform random numbers varying between 0 and 1, v_i^{ite} & v_i^{ite+1} are the existing velocities of the ith particle at iterations (ite) & (ite + 1), respectively, c_1 & c_2 are acceleration coefficients equal to 2, $x_{gbest_i^{ite}}$ is the best position found by all particles till iteration (ite), i.e., Global best, $x_{best_i^{ite}}$ is the best position of the ith particle up to iteration (ite), i.e., Local best, and x_i^{ite} & x_i^{ite+1} are the existing positions of the ith particle at iterations (ite) & (ite + 1), successively.

- Checking the convergence of the S-O framework: Finally, the convergence criterion was set as the relative change in optimal solutions for several successive iterations

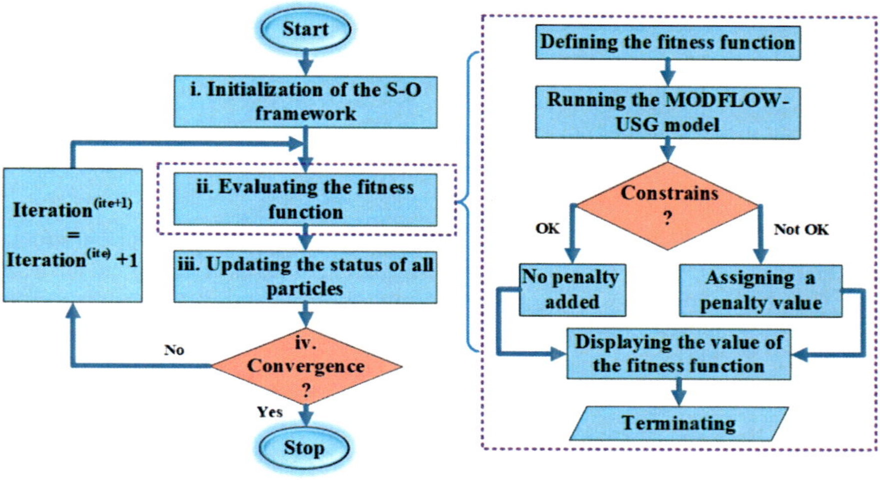

Fig. 4. A flowchart of the proposed simulation-optimization (S-O) framework.

should be less than 0.0001 [7]. If the convergence condition was met, the model would stop. Otherwise, the same procedure of the solution was repeated successively until convergence occurred and the optimum solution was obtained.

3 Results and Discussion

3.1 Calibrated Simulation Model

To assess the accuracy of the simulation model of the Wadi El-Natrun basin after calibration, calculated groundwater heads were drawn against observed ones from 34 observation wells, as displayed in Fig. 5 for both steady and unsteady cases. It was obvious from the figure that all the scatter points were situated very close to the line of the perfect agreement for the two cases. Hence, a good visual match was attained between the calculated and observed heads. In addition, the mean residual between any calculated and observed head did not exceed approximately 0.30 m, which was less than the mean residual value of about 0.85 m obtained by Hao et al. (2018) [5] for calibrating the simulation model of the Beijing plain, China.

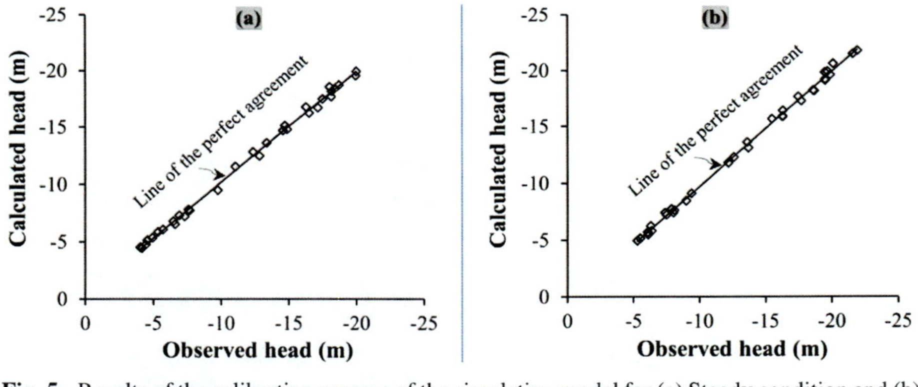

Fig. 5. Results of the calibration process of the simulation model for (a) Steady condition and (b) Unsteady condition.

A further quantitative assessment of the calibration error related to the current simulation model was carried out using different statistical indices for both steady and unsteady states. The statistical indices could give an accurate realization of the error resulting from different models as reported elsewhere [24–28]. The used indices were coefficient of determination (R^2), Nash-Sutcliffe efficiency (NSE), root mean square error (RMSE), and mean absolute error (MAE), as given in Table 1. It was evident from the table that R^2 and NSE, for the two calibration cases approached unity. This value represented a high accuracy for calibrated simulation models, as mentioned by Abd Elmaboud et al. (2021) [7]. Besides, RMSE was equal to about 0.35 m while MAE reached around 0.30 m for the two calibration states. The results obtained here were reasonable compared to the

RMSE and MAE of approximately 0.50 m and 0.30 m, respectively, attained by Kamali and Niksokhan (2017) [11] for the calibrated simulation model of the Esfahan-Borkar area, Iran.

Table 1. Quantitative assessment of the calibration process related to the simulation model of the Wadi El-Natrun basin.

Statistical index	Abbreviation	Formula*	Steady-state calibration	Unsteady state calibration
Coefficient of determination	R^2	$\frac{\sum_{i=1}^{N^o}\left[(h_i^o-h_m^o)(h_i^C-h_m^C)\right]^2}{\sum_{i=1}^{N^o}(h_i^o-h_m^o)^2\sum_{i=1}^{N^o}(h_i^C-h_m^C)^2}$	0.9974	0.9979
Nash-Sutcliffe efficiency	NSE	$1-\frac{\sum_{i=1}^{N^o}(h_i^o-h_i^C)^2}{\sum_{i=1}^{N^o}(h_i^o-h_m^o)^2}$	0.9961	0.9956
Root mean square error	RMSE	$\sqrt{\frac{1}{N^o}\sum_{i=1}^{N^o}(h_i^o-h_i^C)^2}$	0.3417	0.3715
Mean absolute error	MAE	$\frac{\sum_{i=1}^{N^o}\lvert h_i^o-h_i^C\rvert}{N^o}$	0.2920	0.3250

*Note: h_i^o & h_i^C are the observed and calculated groundwater heads at each observation well, h_m^o & h_m^C are the means of the observed and calculated groundwater heads, and N^o is the number of observation wells.

Overall, the simulation model of the Wadi El-Natrun basin exhibited reliable calibration results based on visual and quantitative investigations. Therefore, it could be adopted for future predictions associated with groundwater management within the study area.

3.2 The Output of the Validated Simulation Model

The analysis of the flow budget related to the transient simulation model of the Wadi El-Natrun basin showed that the existing total pumping rate would reach 227000 m^3/day (Base value) at the end of 2020. By increasing the base value by 25%, i.e., 283750 m^3/day (Scenario No. 1) and 50%, i.e., 340500 m^3/day (Scenario No. 2), the results of the simulation model were presented in Fig. 6 after running till December 2030. According to the figure, the predicted groundwater heads for scenarios Nos. 1 and 2 showed that the general groundwater flow directions had not changed and took the same trend of initial heads associated with previous years, i.e., from the northeast to the southwest. However, increasing the pumping discharges using the two development scenarios resulted in a considerable drawdown in groundwater levels in all parts of the model, especially the southwestern part. In addition, the spaces between the contour lines were very close at the beginning of the flow and became wider towards the southwestern portion of the study area. Besides, the contour lines were observed to significantly migrate towards the northeastern direction. Scenario No. 2 was considered the worst condition, causing the contour line of −19 m, which was located in the southwestern area, to move towards the central part of the basin. Similarly, Liu et al. (2018) [29] described the drastic effect of intensive extraction of groundwater on the hydraulic levels in the Subei Lake watershed, in China.

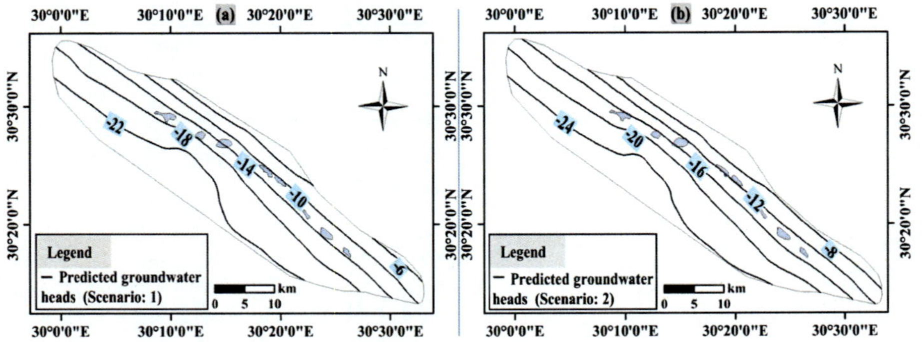

Fig. 6. Predicted groundwater heads for different scenarios: (a) Scenario No. 1 and (b) Scenario No. 2.

For the quantitative realization of the changes in groundwater levels in the basin under study, box plots were drawn, as illustrated in Fig. 7 for the two management scenarios. Box plots could give useful information about the extra drawdowns in groundwater heads at observation wells as reported elsewhere [30]. Referring to Fig. 6, the maximum extra drawdowns would reach 2.80 m and 5.30 m, whereas the minimum values were expected to be 0.39 m and 0.80 m, and the median values were anticipated to be 1.10 m and 2.09 m for scenarios Nos. 1 and 2, respectively. Under these critical conditions of large values of drawdown in some areas, an alternative pumping strategy should be implemented. This strategy ought to regularize the pumping rates of groundwater to achieve acceptable drawdown rates for sustainable development of the study basin.

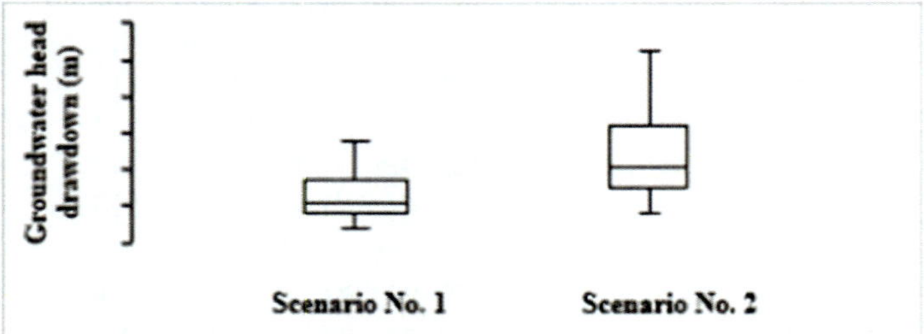

Fig. 7. Statistics of drawdown values in groundwater heads for different scenarios.

3.3 Simulation-Optimization (S-O) Results

The proposed S-O framework was applied in the study basin with a sample size equal to 1000 particles, which was considered the best value as reported by Piotrowski et al. (2020) [31]. This model was used to solve two separate optimization problems, namely cases Nos.1 and 2, based on the statistics attained from box plots. The results were depicted in Fig. 8 after running the model to adjust the drawdown values in groundwater heads at all wells to be less than or equal to 1.10 m and 2.09 m (The median values) for cases Nos. 1 and 2, respectively. It was evident from the figure that the model converged after 1825 iterations for case No. 1, while it took 1843 iterations for case No. 2 until convergence. Besides, it was found that the optimum overall pumping rates were equal to 266556 m^3/day (About 117% of the base value) and 307413 m^3/day (Around 135% of the base value), with a discharge limit for each well to be within the ranges of (69.6–1589.2) m^3/day and (89.1–2078.4) m^3/day for cases Nos. 1 and 2, successively.

These findings could be useful for the optimal exploitation of groundwater resources in the study basin and represent a guide for future horizontal expansions. In addition, the exhibited S-O methodology could be applied to comparable basins to avoid the depletion of aquifers globally. Promising results were also obtained by Abd Elmaboud et al. (2021) [7] using the integration between the MODFLOW-USG model and the PSO algorithm to calibrate the numerical model of a case study in the east of the Nile Delta, Egypt. Therefore, the suggested S-O framework could be further developed to solve multi-objective optimization problems of groundwater.

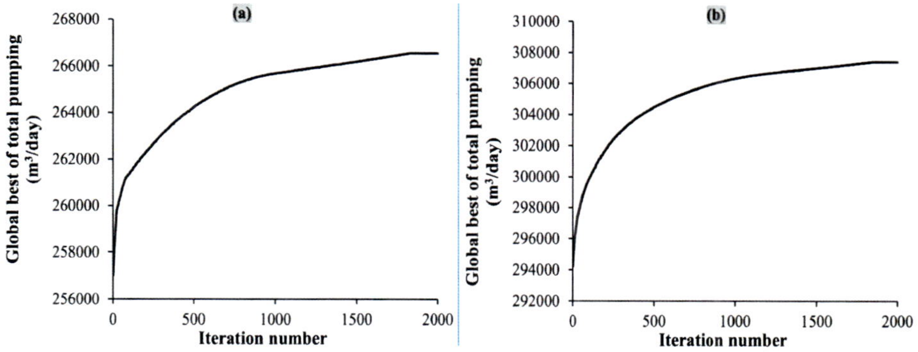

Fig. 8. Optimization results for different cases: (a) Case No. 1, and (b) Case No. 2.

4 Meeting Sustainable Development Goals (SDGs) Through the Current S-O Framework

The current study findings revealed the role of the existing S-O framework to get several benefits in the Wadi El-Natrun basin, including (i) optimal exploitation of groundwater resources, (ii) avoiding the depletion of the Pliocene aquifer, and (iii) guide for future horizontal expansions. These benefits can be correlated to various targets of SDGs with several environmental, economic, and social perspectives.

4.1 Environmental-Based SDGs

The optimal values of groundwater discharges obtained from the proposed S-O framework can lead to positive environmental impacts on the entire ecosystem. In particular, certain limits of withdrawals from the existing Pliocene aquifer can guarantee a sustainable supply of groundwater, viz., Target 6.4 Increase water use efficiency. In addition, this aquifer can be protected from further depletion. As a result, its lifespan can be enhanced to comply with Target 6.6 Restore water-related ecosystems.

4.2 Economic-Based SDGs

The sustainability of groundwater as a feeding source can promote economic growth and industrial development. For instance, keeping groundwater at acceptable levels can help farmers there to maintain the crops' productivity and their associated incomes. Besides, the storage of southern lakes can be preserved to support individual enterprises, especially the salt industry (Target 8.1 Sustain income growth for individuals). Moreover, different existing factories, including food and chemical industries can guarantee their supply of water. In addition, any increase in the cost of groundwater withdrawal can be eliminated to match Target 9.4 Ensuring sustainable industries.

4.3 Social-Based SDGs

Regularizing the pumping rates from groundwater wells can promote the concept of social equality. Every drilled well should have a certain optimal pumping capacity to match Target 10.3 Eliminate inequalities among individuals. Moreover, exposure to water shortage can be prevented. In addition, any losses in people's possessions due to severe land deformations can be averted, i.e., Target 1.5 Reduce the facing of shocks and disasters.

5 Conclusions

A suggested S-O framework was developed for the sustainable use and supply of groundwater in water-stressed regions, and it was applied in a case study, namely the Wadi El-Natrun basin, Egypt. The results of the calibrated simulation model showed that there were good visual and quantitative agreements between the calculated and observed groundwater heads. In addition, the mean residual between those heads, NSE, and RMSE was about 0.3 m, 1, and 0.35 m. By the end of the transient calibration, the water budget analysis revealed that the existing total pumping rate was equal to 227000 m^3/day (Base value). Increasing the base value by 25% (Scenario No. 1) and 50% (Scenario No. 2) might cause a remarkable deterioration in groundwater heads all over the basin under study, especially in the southwestern zone. Besides, the maximum and minimum extra drawdown values in groundwater heads were predicted to be 0.39 m and 2.80 m for scenario No. 1, while they would reach 0.80 m and 5.30 m for scenario No. 2 by 2030. The integrated S-O technique deduced that the optimum overall extraction rates should be 266556 m^3/day (Roughly 117% of the base value) and 307413 m^3/day (Approximately

135% of the base value) to guarantee that the drawdown in heads would not exceed 1.10 and 2.09 (The median values) at the location of pumping wells, respectively. This study revealed that the S-O findings may lead to matching SDGs relevant to affordable water, economic growth, sustainable industries, and reduced inequalities.

The research outputs can be adopted by decision-makers for future expansion activities to avoid further over-exploitation of groundwater in the study basin. Also, the developed S-O framework could be applied in similar basins worldwide to get the best strategies for optimal groundwater management. It was recommended to enhance this model to handle multi-objective groundwater management studies.

Acknowledgments. The first author is very grateful to the Egyptian Ministry of Higher Education (MoHE) for providing financial support in the form of a Ph.D. scholarship. Also, thanks to Japan International Cooperation Agency-JICA for providing all facilities, and equipment to accomplish this research.

References

1. Saqr, A.M., Nasr, M., Fujii, M., Yoshimura, C., Ibrahim, M.G.: Monitoring of agricultural expansion using hybrid classification method in southwestern fringes of Wadi El-Natrun, Egypt: an appraisal for sustainable development. In: Ujikawa, K., Ishiwatari, M., Hullebusch, E.v. (eds.) Environment and Sustainable Development. ACESD 2022. ESE. Springer, Singapore (2023, Forthcoming)
2. Mansour, M.M., Ibrahim, M.G., Fujii, M., Nasr, M.: Recent applications of flash flood hazard assessment techniques: case studies from Egypt and Saudi Arabia. Adv. Eng. Forum **47**, 101–110 (2022). https://doi.org/10.4028/p-03z404
3. Mansour, M.M., Ellayn, A.F., Helal, E., Rashwan, I.M.H., Sobieh, M.F.: Delaying solute transport through the soil using unequal double sheet piles with a surface floor. Ain Shams Eng. J. **9**(4), 3399–3409 (2018). https://doi.org/10.1016/j.asej.2018.10.003
4. Saqr, A.M., Ibrahim, M.G., Fujii, M., Nasr, M.: Sustainable development goals (SDGs) associated with groundwater over-exploitation vulnerability: geographic information system-based multi-criteria decision analysis. Nat. Resour. Res. **30**(6), 4255–4276 (2021). https://doi.org/10.1007/s11053-021-09945-y
5. Hao, Q., Shao, J., Cui, Y., Zhang, Q., Huang, L.: Optimization of groundwater artificial recharge systems using a genetic algorithm: a case study in Beijing. China. Hydrogeol. J. **26**(5), 1749–1761 (2018)
6. Saqr, A.M., Ibrahim, M.G., Fujii, M., Nasr, M.: Simulation-optimization modeling techniques for groundwater management and sustainability: a critical review. Adv. Eng. Forum **47**, 89–100 (2022). https://doi.org/10.4028/p-5011j1
7. Abd Elmaboud, M.E., Abdel-Gawad, H.A., El-Alfy, K.S., Ezzeldin, M.M.: Estimation of groundwater recharge using simulation-optimization model and cascade forward ANN at East Nile Delta aquifer, Egypt. J. Hydrol. Region. Stud. **34**, 100784 (2021)
8. Panday, S., Huyakorn, P.S., Therrien, R., Nichols, R.L.: MODFLOW–USG version 1: an unstructured grid version of MODFLOW for simulating groundwater flow. Adv. Water Resour. **25**(5), 497–511 (2013)
9. Ezzeldin, M.M., El Alfy, K.S., Gawad, H.A., Abd Elmaboud, M.E.: Comparison between structured and unstructured MODFLOW for simulating groundwater flow in three-dimensional multilayer quaternary aquifer of East Nile Delta, Egypt. Hydrol. Curr. Res. **9**(1), 297 (2018)

10. Eberhart, R., Kennedy, J.: A new optimizer using particle swarm theory. In: Proceeding of the IEEE Symposium on Micro Machine and Human Science, Nagoys, Japan, pp. 39–43 (1995)
11. Kamali, A., Niksokhan, M.H.: Multi-objective optimization for sustainable groundwater management by developing of coupled quantity-quality simulation-optimization model. J. Hydroinf. **19**(6), 973–992 (2017)
12. Matott, L.S., Rabideau, A.J., Craig, J.R.: Pump-and-treat optimization using analytic element method flow models. Adv. Water Resour. **29**(5), 760–775 (2006)
13. El-Ghandour, H.A., Elsaid, A.: Groundwater management using a new coupled model of flow analytical solution and particle swarm optimization. Int. J. Water Resourc. Environ. Eng. **5**(1), 1–11 (2013)
14. El-Abd, E.: Geological impact on the water bearing formations in the area Southwest Nile Delta, Egypt. Ph.D. thesis, Faculty of Science, El-Minoufia University (2005)
15. Obaid, A.: Management of an aquifer and identification of the hydraulic parameters using artificial neural networks. M.Sc. thesis, Faculty of Engineering, El-Mansoura University (2015)
16. USGS Homepage. https://earthexplorer.usgs.gov/
17. Switzman, H.R.: The impact of environmental change and water conservation on dryland groundwater resources in northern Egypt: modeling aquifer response using sparse data. M.Sc. thesis, Graduate Studies, McMaster University (2013)
18. El-Osta, M., Hussein, H., Tomas, K.: Numerical simulation of groundwater flow and vulnerability in Wadi El-Natrun depression and vicinities, west Nile Delta, Egypt. J. Geol. Soc. India **92**(2), 235–247 (2018)
19. Ammar, A.: Geophysical studies on groundwater reservoir in Wadi El-Natrun area, western desert, Egypt. M.Sc. thesis, Ain Shams University (2002)
20. El-Boghdady, M.: Integrated hydrogeophysical and environmental studies on groundwater aquifers in Wadi El-Natrun area, Egypt. Ph.D. thesis, Ain Shams University (2017)
21. FAO: Food and agriculture organization of the United Nations database. FAO database (2015). https://wapor.apps.fao.org/catalog/WAPOR_2/1
22. Sedki, A., Ouazar, D.: Swarm intelligence for groundwater management optimization. J. Hydroinf. **13**(3), 520–532 (2011)
23. Zhu, H., Zheng, C., Hu, X., Li, X.: Adaptive PSO using random inertia weight and its application in UAV path planning. In: Seventh International Symposium on Instrumentation and Control Technology, pp. 712–814 (2008)
24. Zidan, A., Abdalla, M., Khalaf, S., Saqr, A.M.: Kinetic energy and momentum coefficients for Egyptian irrigation canals. Mansoura Eng. J. **41**(1), 1–16 (2016). https://doi.org/10.21608/bfemu.2020.99368
25. Zidan, A., Abdalla, M., Khalaf, S., Saqr, A.M.: Resistance equations for Egyptian irrigation canals (case study: 'Dakahliya Governorate'). Int. Water Technol. J. **7**(2), 73–90 (2017). https://www.researchgate.net/publication/362876063_Resistance_equations_for_Egyptian_irrigation_canals_Case_study_Dakahliya_Governorate
26. Zidan, A., Abdalla, M., Khalaf, S., Saqr, A.M.: Regime equations for Egyptian irrigation canals (case study: 'Dakahliya Governorate'). Int. Water Technol. J. **8**(4), 129–141 (2018). https://www.researchgate.net/publication/363196196_Regime_equations_for_Egyptian_irrigation_canals_Case_study_Dakahliya_Governorate
27. Mansour, M.M., Ibrahim, M.G., Fujii, M., Nasr, M.: Sustainable development goals (SDGs) associated with flash flood hazard mapping and management measures through morphometric evaluation. Geocarto Int. (2022). https://doi.org/10.1080/10106049.2022.2046868
28. Mansour, M.M., Nasr, M., Fujii, M., Yoshimura, C., Ibrahim, M.G.: Quantification of flash flood run off volume using morphometric parameters towards sustainability. In: Ujikawa, K., Ishiwatari, M., Hullebusch, E.v. (eds.) Environment and Sustainable Development. ACESD 2022. ESE. Springer, Singapore (2023, Forthcoming)

29. Liu, F., et al.: Predicting the impact of heavy groundwater pumping on groundwater and ecological environment in the Subei Lake basin, Ordos energy base, northwestern China. Hydrol. Res. **49**(4), 1156–1171 (2018)
30. Das, N., Sutradhar, S., Ghosh, R., Mondal, P., Islam, S.: The response of groundwater to multiple concerning drivers and its future: a study on Birbhum district, West Bengal, India. Appl. Water Sci. **11**(4), 1–22 (2021)
31. Piotrowski, A.P., Napiorkowski, J.J., Piotrowska, A.E.: Population size in particle swarm optimization. Swarm Evol. Comput. **58**, 100718 (2020)

Estimating VKT on Rural Roads for Carbon Emission Estimation

Peng Zhang[1](✉), Lifeng Xing[1], Lukai Zhang[1], and Zhenyu Li[2]

[1] Transport Planning and Research Institute, Ministry of Transport, Beijing, China
zhangpeng@tpri.org.cn
[2] Zhonglugang Engineering Consulting Co., Ltd., Beijing, China

Abstract. To estimate VKT on rural roads, an estimation method based on the fusion of satellite remote sensing images and traffic detection data was presented. The ratio of vehicle densities between different categories of highways was obtained with satellite remote sensing data. The speed of rural highways was estimated using the data from traffic detectors and the design speeds of highways. Taking the traffic volume of high-level highway obtained by the traffic detectors as the benchmark, the average traffic volume and VKT of rural highways were estimated according to the ratio of vehicle density and the ratio of vehicle speed of rural highways to those of high-level highways. We tested the method with data from random sampling areas. The results show that the traffic volume and VKT of rural roads in China are 187 vehicles/day and 8.35 × 108 vehicle·kilometers/day, respectively. The proposed method does not dependent on the image scale, so it is also applicable to unmanned aerial vehicle photography.

Keywords: Carbon peak & carbon neutrality · Rural road · Data fusion · Vehicle-kilometers traveled · Satellite remote sensing

1 Introduction

To achieve the goal of carbon peaking and carbon neutrality in transportation, it is necessary to master the current carbon emission amount of each transportation mode. Carbon emission estimation methods are divided into two categories: energy consumption based and transportation activities based. The energy consumption method calculates carbon emissions based on fuel consumption, which has less calculation demand and less error in conversion between calculation steps. The method based on traffic activities calculates carbon emissions according to traffic and transportation intensity, and carbon emissions can be predicted according to the prediction results of traffic activities. In addition, in the calculation of road transportation carbon emissions, the impact of agricultural machinery refueling, smuggled fuel and fuel sales without registration on energy consumption can be avoided with activities based method. In the practical application, both methods are used and mutually verified.

For the traffic activities based method, the carbon emissions of railway, water and air transportation are calculated according to the passenger/freight ton-kilometers and

X. Chen (Ed.): ICESE 2022, ESE, pp. 74–86, 2023.

vehicle emission factors. The transportation statistics are facing commercial transportation. There is no non-commercial transportation for these three modes of transportation. It is convenient to calculate carbon emissions according the ton-kilometers obtained from transportation statistics. For road transportation, carbon emissions are calculated according to vehicle-kilometers traveled (VKT) and vehicle emission factors [1]. Due to the existence of a large number of non-commercial vehicles such as private cars and non-commercial trucks in the roads, the VKT of all vehicles is difficult to obtain from transportation statistics. At present, China makes transportation statistics in term of the registration place of transportation enterprises. No matter where the vehicle is driven, the VKT is included in the registered place of the enterprise, rather than the actual place of transportation happened. Therefore, it is difficult to obtain the carbon emissions of a region according to the VKT from statistics. The estimation of VKT is a key problem in the estimation of transportation carbon emissions.

Besides the statistics, VKT can be obtained through estimation. In the early years, the coverage of road traffic detectors was low. VKT was mainly calculated using data other than road traffic detection. Hossain [2] calculated VKT according to fuel sales and vehicle fuel consumption. Ferreira [3] obtained the annual mileage according to the difference of the odometer data of the annual vehicle inspection, and calculates VKT in combination with the vehicle ownership. Greene [4] obtained personal driving distance through survey, and calculated VKT in combination with the number of drivers. Blei [5] obtained the travel OD according to the survey, got the travel trajectory using traffic assignment model, and then calculated VKT. There are some developments of method based on travel survey. Lead [6] used the online questionnaire based on Google API, registered the origin and destination of the respondents, and calculated VKT. Albertini [7] used statistical models to improve the accuracy of VKT estimation.

With the increase of the coverage of road traffic detectors, the traffic volume based method [8] has been widely used. It takes the product of traffic volume and highway mileage as the estimation result of VKT. Considering the differences of traffic conditions in different road segments, the traffic volume based method has been optimized [9, 10]. Teng [11] analyzed the requirements of VKT estimation on the sampling rate of discontinuous traffic volume detection. Klatko [12] estimated the VKT of the roads without detectors by interpolation method according to the road categories. The development of ITS technologies enriched the method of VKT estimation. Yan [13] used the freeway charging data to obtain the trajectory of vehicles and calculate the VKT of highways. Based on the global positioning data, nasri [14], liu [15] and fan [16] matched the vehicle position with GIS map to obtain the trajectory and calculate the VKT. Based on various VKT estimates, Huo [17] analyzed the relationship between vehicle age, vehicle type and other factors and VKT to correct the VKT estimates. For rural areas, Yao [18] sampled the VKT of hundreds of rural vehicles in Henan province. But the estimation results include the VKT of agricultural vehicles off road, and the VKT on road cannot be separated.

Roads in China are divided into urban roads and highways. According to their functions, highways are divided into four categories, including freeways, ordinary state highways, ordinary provincial highways and rural highways. The first three are called arterial highway. Rural highways consist of county roads, township roads and village roads. By the end of 2021, the mileage of rural highways in China was 4.46 million kilometers, accounting for 84.5% of the total mileage of highways. Rural highways are large in scale, complex in natural conditions, difficult in power supply and maintenance, lack of traffic detectors.

It is difficult to obtain the basic data required by the methods based on traffic volume. Because it is difficult to separate the on road and the off road traffic in rural areas. The fuel consumption based method is difficult to apply. At present, the vehicles that can provide global positioning data only include trucks with a total weight of more than 12 tons and some other commercial vehicles, so the method based on vehicle trajectory is difficult to apply. The rural area is vast, and the budget required to carry out the traffic survey of nationwide rural highways is too large, the survey based method is also difficult to apply. Rural highways are the difficulty in VKT estimation, and also one of the main technical bottlenecks in transportation carbon emission estimation.

By the end of 2021, China has built more than 11000 automated traffic detection stations in highways, covering all 763 thousand kilometers of arterial highway and a small number of county roads, and collecting real-time data of traffic volume, vehicle speed, vehicle types and axle loads. Meanwhile, China has more than 200 remote sensing satellites, of which the surface image resolution of optical remote sensing satellites reaches decimeter level. This paper fused the data of satellite remote sensing and automated traffic detection to solve the problem of VKT estimation on rural roads. This paper is organized as follows: the second section gives the VKT estimation method, the third section gives the calculation results, and the conclusions are presented in the last section.

2 Methodology

2.1 Estimating Average Traffic Volume on Rural Highways

According to the traffic volume and the average speed of vehicles, the vehicle density on the highway can be calculated, as shown in Eq. (1).

$$\rho = \frac{c}{24v} \tag{1}$$

where ρ is the density of vehicles (vehicle/km), c is the traffic volume (vehicle/day), c is the average speed of vehicles (km/h). Equation (1) can be converted into Eq. (2).

$$c = 24v\rho \tag{2}$$

Let c_i, ρ_i, v_i, L_i represent the traffic volume, vehicle density, vehicle speed and mileage of different highway categories, $i = 1, 2, 3, 4$ correspond to the freeways, ordinary state highways, ordinary provincial highways and rural highways, respectively. Then Eq. (3) can be obtained from Eq. (2).

$$\frac{c_4}{c_i} = \frac{\rho_4 v_4}{\rho_i v_i} \tag{3}$$

where $i = 1, 2, 3$. Then Eq. (4) can be obtained.

$$c_4 = c_i \times \frac{\rho_4}{\rho_i} \times \frac{v_4}{v_i} \tag{4}$$

where c_i is the mean of the traffic volume of all traffic detection stations with the same highway category. ρ_4 / ρ_i is the ratio of vehicle density of rural highways to that of the ith category of highways, as given in Sect. 2.2. v_4 / v_i is the ratio of vehicle speed of rural highways to that of the ith category of highways, as given in Sect. 2.3.

2.2 Ratio of Vehicle Density of Rural Highways to that of High-Level Highways

For large-scale rural roads, the research method of sampling survey was adopted. A number of areas were randomly selected as sampling areas, and the satellite remote sensing image data of the sampling areas were retrieved. The resolution of remote sensing data should meet the requirements of identifying the vehicles on the roads. The remote sensing data was used to obtain the length of roads and the number of vehicles in each category of roads in sampling area. Then ρ_4 / ρ_i can be obtained from Eq. (5).

$$\frac{\rho_4}{\rho_i} = \frac{\sum_{j=1}^{N} n_{4,j}}{\sum_{j=1}^{N} l_{4,j}} \Bigg/ \frac{\sum_{j=1}^{N} n_{i,j}}{\sum_{j=1}^{N} l_{i,j}} \tag{5}$$

where $n_{i,j}$ and $l_{i,j}$ are the number of vehicles and road mileage in the ith category of highways and the jth sampling area, respectively, N is the number of sampling areas. $n_{i,j}$ and $l_{i,j}$ can be obtained by computer image recognition or artificial counting. If a category of highway did not exist in the sampling area, the number of vehicles and road mileage were set to zero.

2.3 Ratio of Vehicle Velocity of Rural Highways to that of High-Level Highways

Let $\overline{v}_{4,k}$, $L_{4,k}$, $v_{4,k}$ represent the design speed, mileage and vehicle speed of different categories of rural roads, where $k = 1, 2, 3$ correspond to the county roads, township roads and village roads, respectively. $v_{4,1}$ can be obtained from the traffic detectors installed on the county roads. There is a great correlation between the traffic states of various categories of rural roads in a small area. Taking the speed of county roads as the benchmark, the speed of township roads and the speed of village roads were calculated according to the ratio between the design speeds of county roads, township roads and village roads, as shown in Eq. (6).

$$v_{4,k} = \frac{\overline{v}_{4,k} v_{4,1}}{\overline{v}_{4,1}} \tag{6}$$

Then the vehicle speed of rural highways can be obtained from Eq. (7).

$$v_4 = \frac{\sum_{k=1}^{3} L_{4,k} v_{4,k}}{\sum_{k=1}^{3} L_{4,k}} \tag{7}$$

Traffic detectors were installed on freeways, ordinary state highways and ordinary provincial highways. $v_i, i = 1, 2, 3$ can be obtained from automated detection data. Then v_4 / v_i can be calculated.

2.4 Estimating VKT on Rural Highways

Substituting ρ_4 / ρ_i obtained in Sect. 2.2 and v_4 / v_i obtained in Sect. 2.3 into Eq. (4), we obtained c_4. Substituting c_4 into Eq. (8), we got the VKT of rural highways.

$$y_4 = c_4 L_4 \tag{8}$$

where y_4 is the VKT on rural highways. The VKT of different types of vehicles can be calculated by using the detection and recognition data of different types of vehicles according to Eqs. (4) to (8).

3 Estimation Results

3.1 Data Specification

We carried out random sampling in China. Data sources: remote sensing images and automated traffic detection data. Spatial scope: nationwide highway network in China. Size of sampling area: 5 to 20 m^2. Sampling size: 35 areas. Vehicle recognition method of remote sensing image: artificial recognition. Considering the functions of different categories of highways in the national economy and daily social activities, the traffic of ordinary provincial highways are more relevant to that of rural highways than freeways and ordinary state highways. Therefore, the ordinary provincial highways were taken as the benchmarks to calculate the VKT of rural highways. The number of vehicles and mileage of the ordinary provincial highways and rural highways in the sampling area are shown in Table 1.

3.2 VKT Estimation Result

ρ_4 / ρ_3 can be obtained by substituting the data in Table 1 into Eq. (5).

$$\begin{aligned}\frac{\rho_4}{\rho_3} &= \frac{\sum_{j=1}^{35} n_{4,j}}{\sum_{j=1}^{35} l_{4,j}} \Bigg/ \frac{\sum_{j=1}^{35} n_{3,j}}{\sum_{j=1}^{35} l_{3,j}} \\ &= 0.062\end{aligned}$$

According to the detector data, $c_3 = 6749$ vehicles/day, $v_3 = 55.3$ km/h. Table 2 shows the design speeds of county roads, township roads and village roads and the mileages.

According to the data of 494 automated traffic detection stations deployed in county roads, we got $v_{4,1} = 52.2$ km/h. Substituting $v_{4,1}$ and $\overline{v}_{4,k}$ in Table 2 into Eq. (6)

Table 1. Vehicle recognition results of satellite remote sensing images.

	Location	Route code	Ordinary provincial highways		Rural highways	
			Length (km)	Vehicle number	Length (km)	Vehicle number
1	Changfu District, Liaocheng City, Shandong Province	S325	1.35	25	7.26	12
2	Xiagezhuang Town, Pinggu District, Beijing	S314	1.02	8	7.39	2
3	Xindu District, Xingtai City, Hebei Province	S322	1.14	4	3.30	1
4	Qujiang District, Shaoguan City, Guangdong Province	S106	0.49	5	4.25	4
5	Qingtongxia City, Wuzhong City, Ningxia Hui Autonomous Region	S103	1.10	2	11.63	2
6	Bayingolin Mongolian Autonomous Prefecture of Xinjiang Uygur Autonomous Region	S307	1.31	3	22.56	8
7	Datong Hui and Tu Autonomous County, Xining City, Qinghai Province	S102	0.63	2	8.67	1

(continued)

Table 1. (*continued*)

	Location	Route code	Ordinary provincial highways		Rural highways	
			Length (km)	Vehicle number	Length (km)	Vehicle number
8	Yongjing County, Linxia Hui Autonomous Prefecture, Gansu Province	S105	1.06	6	3.43	2
9	Lushan County, Ya'an City, Sichuan Province	S210	0.30	2	2.89	2
10	Chenggong District, Kunming City, Yunnan Province	S102	0.68	23	4.75	11
11	Linghai City, Jinzhou City, Liaoning Province	S306	0.29	2	1.37	2
12	Nanguan District, Changchun City, Jilin Province	S104	0.9	13	1.93	7
13	Urumqi county, Urumqi City, Xinjiang Autonomous Region	/	0	0	3.02	0
14	Shenzha County, Naqu City, Tibet Autonomous Region	/	0	0	1.96	0

(*continued*)

Table 1. (*continued*)

	Location	Route code	Ordinary provincial highways		Rural highways	
			Length (km)	Vehicle number	Length (km)	Vehicle number
15	Luchun County, Honghe Hani and Yi Autonomous Prefecture, Yunnan Province	/	0	0	1.05	1
16	Lingshan County, Qinzhou City, Guangxi Zhuang Autonomous Region	/	0	0	1.07	0
17	Anxi County, Quanzhou City, Fujian Province	/	0	0	1.61	0
18	Yuhang District, Hangzhou City, Zhejiang Province	/	0	0	1.60	1
19	Nanchang County, Nanchang City, Jiangxi Province	/	0	0	1.61	1
20	Pizhou City, Xuzhou City, Jiangsu Province	/	0	0	2.03	1

(*continued*)

Table 1. (*continued*)

	Location	Route code	Ordinary provincial highways		Rural highways	
			Length (km)	Vehicle number	Length (km)	Vehicle number
21	Nanzhao County, Nanyang City, Henan Province	/	0	0	1.89	0
22	Wuxi County, Chongqing	/	0	0	1.54	0
23	Aba Prefecture, Aba Tibetan and Qiang Autonomous Prefecture, Sichuan Province	/	0	0	1.07	1
24	Weng'an county, Qiannan Buyei and Miao Autonomous Prefecture, Guizhou Province	/	0	0	1.54	2
25	Anding District, Dingxi City, Gansu Province	/	0	0	1.94	0
26	Wulat Middle Banner, Bayannur City, Inner Mongolia Autonomous Region	/	0	0	2.44	0

(*continued*)

Table 1. (*continued*)

	Location	Route code	Ordinary provincial highways		Rural highways	
			Length (km)	Vehicle number	Length (km)	Vehicle number
27	Zhiduo County, Yushu Tibetan Autonomous Prefecture, Qinghai Province	/	0	0	2.66	0
28	Zhen'an County, Shangluo City, Shaanxi Province	/	0	0	1.05	0
29	Lanxi County, Suihua City, Heilongjiang Province	/	0	0	2.5	0
30	Aipu River, Qingpu District, Shanghai	/	0	0	7.3	6
31	Qianshan County, Anqing City, Anhui Province	/	0	0	6.2	4
32	Shuangfeng County, Loudi City, Hunan Province	/	0	0	2.1	1
33	Pingyao County, Jinzhong City, Shanxi Province	/	0	0	1.51	1

(*continued*)

Table 1. (*continued*)

	Location	Route code	Ordinary provincial highways		Rural highways	
			Length (km)	Vehicle number	Length (km)	Vehicle number
34	Hua county, Anyang City, Henan Province	/	0	0	1.48	2
35	Yangxin County, Huangshi City, Hubei Province	/	0	0	0.70	0
Total			10.27	95	130.6	75

Table 2. Design speeds and mileages of different categories of rural highways.

Road category	$\overline{v}_{4,k}$ (km/h)	$L_{4,k}$ ($\times 10^3$ km)
County road	60	580
Township road	30	1198
Village road	20	2422

can obtain $v_{4,2} = 26.1$ km/h and $v_{4,3} = 17.4$ km/h. Then Substituting $v_{4,k}$ and $L_{4,k}$, $k = 1, 2, 3$ in Table 2 into Eq. (7) can obtain $v_4 = 24.7$ km/h.

Then substituting ρ_4 / ρ_3, c_3, v_3 and v_4 into Eq. (4) can get the average traffic volume of rural highways.

$$\begin{aligned} c_4 &= c_3 \times \frac{\rho_4}{\rho_3} \times \frac{v_4}{v_3} \\ &= 187(vechile/day) \end{aligned}$$

By the end of 2021, the mileage of rural highways in China is 4.46 million kilometers. Substituting c_4 and L_4 into Eq. (8) can get the total VKT on rural highways in China.

$$\begin{aligned} y_4 &= c_4 L_4 \\ &= 8.35 \times 10^8 (vehicle \cdot km/day) \end{aligned}$$

4 Conclusions

Due to budget constraints, traffic detectors are always preferentially deployed on high-level highways or highways with large traffic demand. Due to the lack of detection equipment, large scale rural highways have long been a blind spot for traffic surveillance. The proposed method took advantage of the wide coverage of remote sensing image data, calculated the VKT of rural highways through nested analysis of multisource data. The much-needed VKT data of rural traffic for carbon emission estimation were provided. For the heavy workload of artificial image recognition, only 35 samples were selected. In the future, image recognition technologies would be used to improve the efficiency of identifying vehicles in the road. With the increase of sample size, the performance of the proposed method will be improved.

Acknowledgement. This paper was supported by National Natural Science Foundation of China (No. 62173167), Program of Strategy, Plan and Policy of Transport of China (No. 2020-23-2).

References

1. Wang, Y.J., Wang, M.Y., Ji, Z.: Study and analysis on vehicle emission models abroad. Environ. Sustain. Dev. **45**(5), 159–164 (2020)
2. Hossain, A., Gargett, D.: Road vehicle-kilometers travelled estimated from state/territory fuel sales. In: Australasian Transport Research Forum, Adelaide (2011)
3. Ferreira, J., Diao, M., Xu, J.: Estimating the vehicle-miles-traveled implications of alternative metropolitan growth scenarios: a Boston example. Trans. GIS **17**(5), 645–660 (2013)
4. Greene, D.L.: Long-run vehicle travel prediction from demographic trends. Transp. Res. Rec. **1135**, 1–9 (1987)
5. Blei, A., Kawamura, K., Javanmardi, M., et al.: Evaluation methods for estimating vehicle miles traveled with GPS travel survey data. Transp. Res. Rec. **2495**(1), 112–120 (2015)
6. Leard, B., Linn, J., Munnings, C.: Explaining the evolution of passenger vehicle miles traveled in the United States. Energy J. **40**(1), 25–54 (2019)
7. Alberini, A., Burra, L.T., Cirillo, C., et al.: Counting vehicle miles traveled: what can we learn from the NHTS. Transp. Res. Part D: Transp. Environ. **98**(3), 102984 (2021)
8. Highway performance monitoring system field manual. Federal Highway Administration, U.S. Department of Transportation (2010)
9. Yu, L., Qiao, F.X., Liu, J.: VMT estimation model based on traffic counts and link attributes. J. Transp. Syst. Eng. Inf. Technol. **5**(1), 38–46 (2005)
10. Kim, S., Park, D., Heo, T.Y., et al.: Estimating vehicle miles traveled (VMT) in urban areas using regression kriging. J. Adv. Transp. **50**(5), 769–785 (2016)
11. Teng, H., Wang, N.: Estimating vehicle miles traveled combined with ITS data. Transp. Plan. Technol. **34**(8), 777–794 (2011)
12. Klatko, T.J., Agbelie, B.R., Labi, S., et al.: Estimation and prediction of statewide Vehicle Miles Traveled (VMT) by highway category and vehicle classification. Joint Transportation Research Program, Purdue University (2016)
13. Yan, S.Y., Xiao, R.M.: Index characteristics of expressway transportation volume based on toll collection data. J. Traff. Transport. Eng. **18**(6), 112–120 (2018)
14. Nasri, A., Zhang, L., Fan, J., et al.: Advanced vehicle miles traveled estimation methods for non-federal aid system roadways using GPS vehicle trajectory data and statistical power analysis. Transp. Res. Rec. **2673**(11), 296–308 (2019)

15. Liu, W., Hao, L.,Li, F.: Estimation of vehicle miles traveled based on improved ant colony algorithm. SAE Technical Paper (2020)
16. Fan, J., Fu, C., Stewart, K., et al.: Using big GPS trajectory data analytics for vehicle miles traveled estimation. Transport. Res. Part C Emerg. Technol. **103**, 298–307 (2019)
17. Huo, H., Zhang, Q., He, K., et al.: Vehicle-use intensity in China: current status and future trend. Energy Policy **43**, 6–16 (2012)
18. Yao, Z., Huo, H., Zhang, Q., et al.: Gaseous and particulate emissions from rural vehicles in China. Atmos. Environ. **45**(18), 3055–3061 (2011)

Toxic Effects of Chromium (VI) on *Chlorella vulgaris* in Water

Zhouyi Zhou, Yicheng Wu(✉), Binghui Liu, Chen Yu, and Haiyan Fu

College of Environmental Science and Engineering, Xiamen University of Technology, Xiamen, China
ycwu@xmut.edu.cn

Abstract. Considering that Chromium (VI) is highly toxic to living organisms, *Chlorella vulgaris* was used as the test organism to investigate the effects of Cr (VI) at different concentrations (1, 3, 6 and 9 mg/L). The results indicated that the growth and the photosynthetic activity of *C. vulgaris* were inhibited under the environmental stress of Cr (VI) with the all concentrations tested, which significantly enhanced with the increase of exposure time. Although *C. vulgaris* is capable of growth exposure to 1, 3 and 6 mg/L Cr (VI) concentrations, the dry weights of *C. vulgaris* were 92.2%, 84.0% and 66.5% of the control group exposed to 1, 3, 6 mg/L Cr (VI), respectively. The most remarkable inhibition effects observed when *C. vulgaris* exposed to Cr (VI) concentration of 9 mg/L for 4 days, and the inhibition rate of growth, chlorophyll a content, Fv/Fm and ΦPSII were 52%, 65.3%, 34.5% and 44.2%. The results demonstrated that photosystem II was damaged under exposure to Cr (VI) and thus reduce the efficiency of photosynthesis, which affected the growth of *C. vulgaris*.

Keywords: Chromium · *Chlorella vulgaris* · Photosynthetic

1 Introduction

In recent decades, the issues of heavy metal pollution in environment were becoming severe and received a rising concern along with the high-speed development in industry. chromium (Cr) is one of the most frequently detected heavy metal in environment, which was primarily produced by mining, metal smelting, leather tanning, electroplating and other industries [1]. Due to its unable decomposed naturally, Cr would long-time presence in environment and posed serious threats to biology [2, 3]. Compared with Cr (III), Cr (VI) is higher toxicity owing to Cr (VI) more easily diffuses across cell membranes and accumulated in cells [4]. Although there are many physical, chemical, and biological remediation methods of Cr (VI), while not enough to deal with the issues of Cr (VI) pollution in environment [5]. That Cr (VI) concentration continuously increased in aquatic systems has adverse effects to organisms and can cause serious diseases on human such as skin diseases [6], respiratory diseases [7], cancers [1].

Chlorella vulgaris which is majority producer in environment plays a crucial role in maintaining the balance of aquatic ecosystems and frequently utilized for evaluating

X. Chen (Ed.): ICESE 2022, ESE, pp. 87–93, 2023.

ecological risk of contaminants [8]. Additionally, chlorophyll fluorescence characteristics, such as Fv/Fm and ΦPSII, were widely used as probes to monitor the effect of contaminant on the photochemical efficiency of microalgae [9]. In this study, the effects of different concentrations of Cr (VI) stress on biomass and chlorophyll a content of *C. vulgaris* were investigated, and further explored the response of chlorophyll fluorescence characteristics to different concentrations of Cr (VI), which can provide a basis for ecological risk assessment of Cr (VI) in aquatic ecosystems.

2 Materials and Methods

2.1 Chemicals and Reagent

$K_2Cr_2O_7$ (99% purity) obtained from the Sinopharm Chemical Reagent Co., Ltd. (Shanghai, China) and storage solution of Cr (VI) was prepared in ultrapure water.

2.2 Microalgae Cultivation

The *C. vulgaris* samples were purchased from the Yuanquan Biotechnology Co., Ltd. (Hainan, China). The microalgae were activated in 250 mL conical flask with autoclaved BG11 medium, and further cultivated at 3000 lx light intensity and 26 ± 1 °C with 14 h–10 h light-dark photoperiod. All of experimental apparatus was involved sterilization in autoclave with 121 °C for 15 min.

2.3 Experimental Design and Test Methods

Cr (VI) treatment concentrations were 1, 3, 6, and 9 mg/L, which were parallel three times. The changes in the biomass were determined every day for 4 day through spectrophotometer method at 680 nm and further transform into the cell dry weight by the equation: Cell dry weight $= 0.2262 \times$ OD680 $- 0.0167$ ($R^2 = 0.9941$). The chlorophyll fluorescence characteristics (Fv/Fm, ΦPSII) were analyzed by PHYTO-PAM phytoplankton fluorometer [10]. Inhibition rate (%) $= (C_0 - C_t)/C_0$ was utilized to figure up the inhibition rate, where C_0 is the control, C_t is the treatment.

2.4 Statistical Analysis

The diagrams in this paper were achieved by the Origin 9.0. One-way analysis of variance (ANOVA) was applied to assess the difference between groups, which was completed by the SPSS 22.0. Among them, when the value of P below 0.05 was represented the significant difference, and below 0.01 was expressed the extreme significant difference.

3 Experimental Results and Analysis

3.1 Effects of Difference Cr (VI) Concentration on the Growth of Chlorella vulgaris

Effects of different Cr (VI) on dry weight of *C. vulgaris* were shown in Fig. 1a. As shown in Fig. 1a, the growth of *C. vulgaris* was obviously inhibited under the stress of

Cr (VI) with the all concentrations tested ($p < 0.05$). *C. vulgaris* could constantly grow at Cr (VI) concentration of below 3 mg/L, whereas biomass continuously decreased to 0.147 g/L exposed to 9 mg/L Cr (VI). Notably, the dry weight of *C. vulgaris* decreased to 0.193 g/L, while exposed to 6 mg/L Cr (VI) on the second day. *C. vulgaris* were adapted to the stress of Cr (VI) with the extension of stress time, and dry weight increased to 0.203 g/L in the following two days. The growth inhibition rate of *C. vulgaris* was showed in Fig. 1b. After 4d of Cr (VI) exposure, the toxicity of Cr (VI) to *C. vulgaris* constantly increased, the inhibition rate rose to 7.8%, 16.0%, 33.5% and 52% with the increase in Cr (VI) concentration, respectively. The results demonstrated that Cr (VI) possessed the acute toxicity to microalgae, and inhibitory effect on cells significantly correlated with Cr (VI) concentration and exposure time ($p < 0.01$). The results are similar to investigation reported by Zhang [11].

3.2 Effects of Cr (VI) Exposure Concentration on Chlorophyll a Content

A significant difference of chlorophyll a content can be observed between the Cr (VI) exposure groups and the control group ($p < 0.05$) in Fig. 2a. During the whole incubation, chlorophyll a content was significantly inhibited under exposure to Cr (VI) and the inhibition was drastically enhanced with the increase in time (Fig. 2b). The chlorophyll a content in cell of *C. vulgaris* exposed to 1 and 3 mg/L Cr (VI) slightly raised in the first two days and then decreased to 10.46 mg/L and 9.01 mg/L, respectively. Chlorophyll a content drastically decreased from 10.82 mg/L at the beginning to 4.98 mg/L exposed to 9 mg/L Cr (VI) for 4 days, indicated that severe inhibition of chlorophyll a synthesis under high Cr (VI) concentration stress. Compared with the change of biomass, we found that chlorophyll a was more sensitive to Cr (VI). These results showed that chlorophyll a was damaged under exposure of Cr (VI).

3.3 Effects of Cr (VI) on the Photochemical Efficiency of Photosystem II

Effect of pollutants on photosynthetic activity of microalgae can be reflected by the value of Fv/Fm, which generally evaluate the maximum photosynthetic capacity in PSII of microalgae under stress. The effects of Cr (VI) on Fv/Fm value of *C. vulgaris* was shown in Fig. 3a, the value of Fv/Fm increased in first day exposed to 1 and 3 mg/L Cr (VI), and then fell to 0.50 and 0.46, respectively. Additionally, Fv/Fm was rapidly decreased to 0.43 and 0.37 with Cr (VI) concentration increased to 6 and 9 mg/L in four days, respectively.

ΦPSII is an essential parameter to assess the actual photochemical efficiency of PSII reactor center. It can be seen from Fig. 3b that ΦPSII showed a similar trend to chlorophyll a content and Fv/Fm with extreme significant correlation when *C. vulgaris* exposed to Cr (VI). ΦPSII of *C. vulgaris* decreased to 0.15, 0.14, 0.13 and 0.11 with the increase in Cr (VI) concentration in four days, respectively. Such results implied that PSII reactor center was damaged when exposure to Cr (VI), and the degree of damage enhances with exposure Cr (VI) concentration.

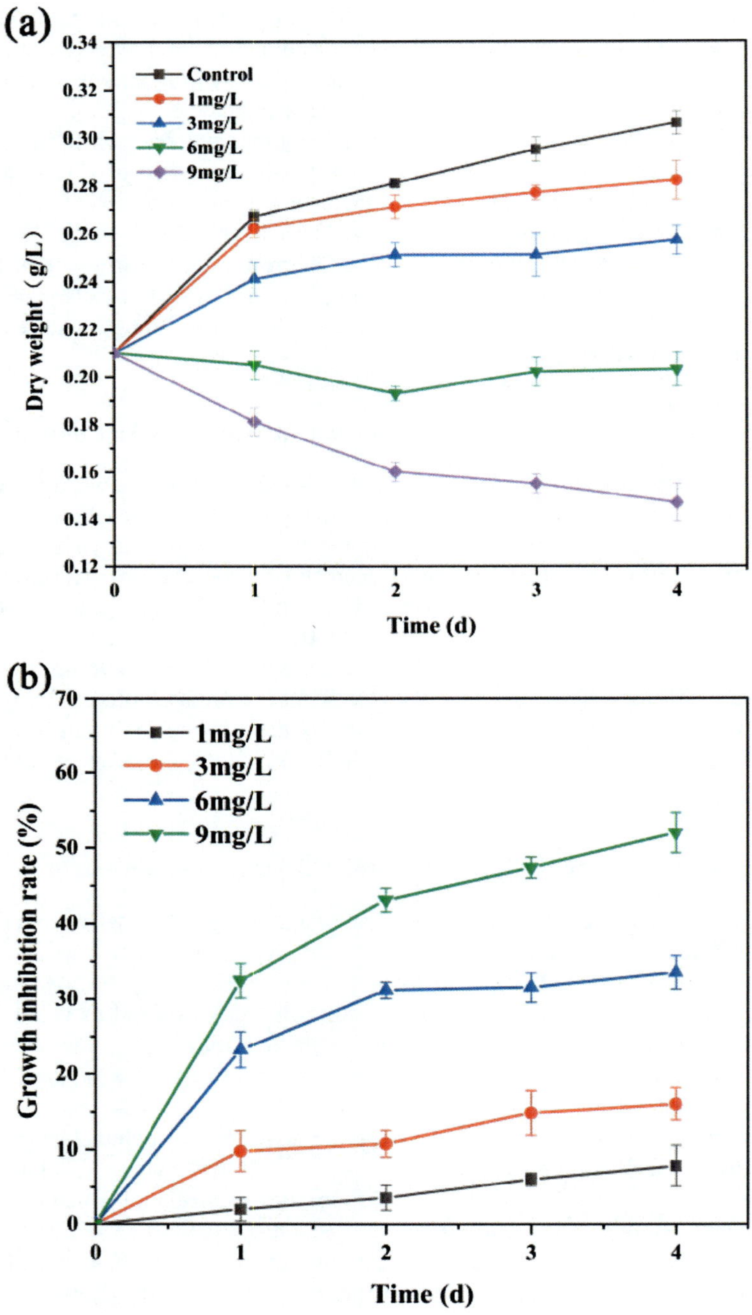

Fig. 1. Effects of different Cr (VI) on dry weight of *Chlorella vulgaris*

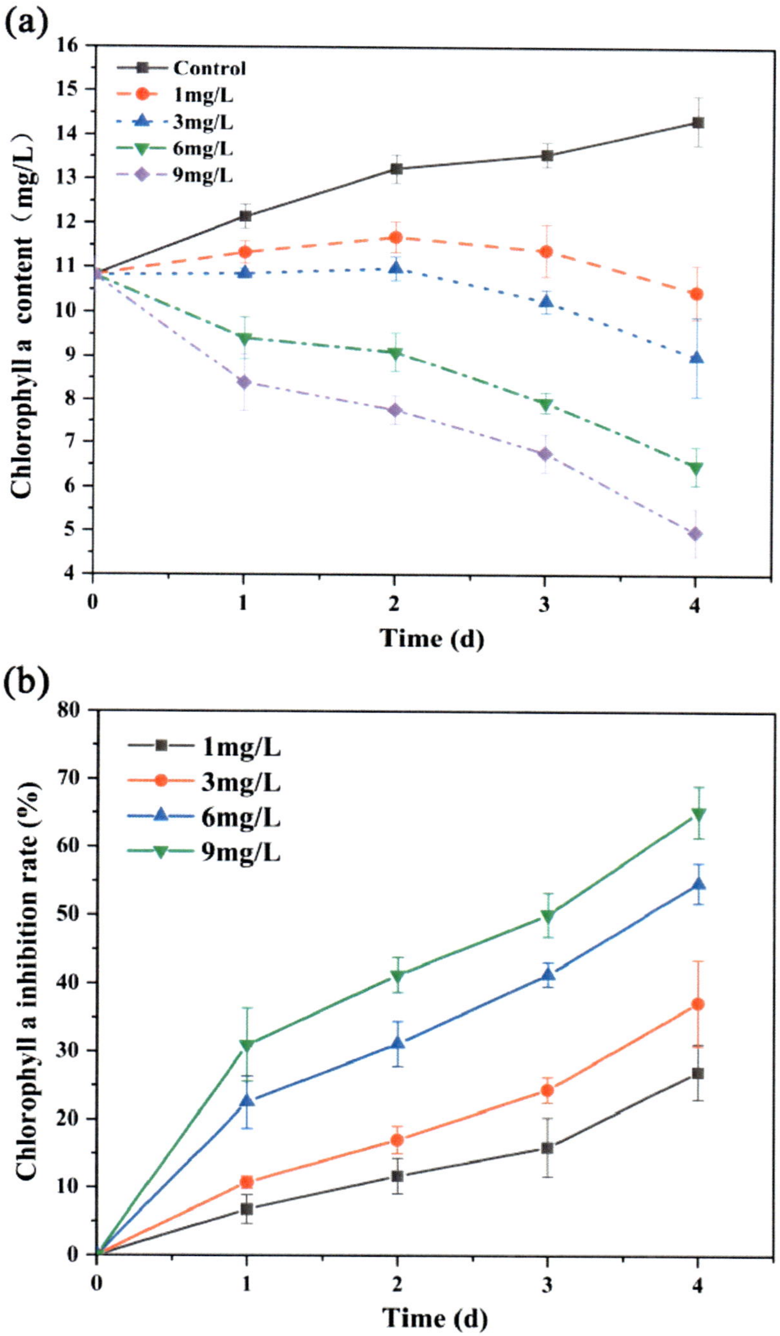

Fig. 2. Effects of Cr (VI) on the chlorophyll a content of *Chlorella vulgaris*

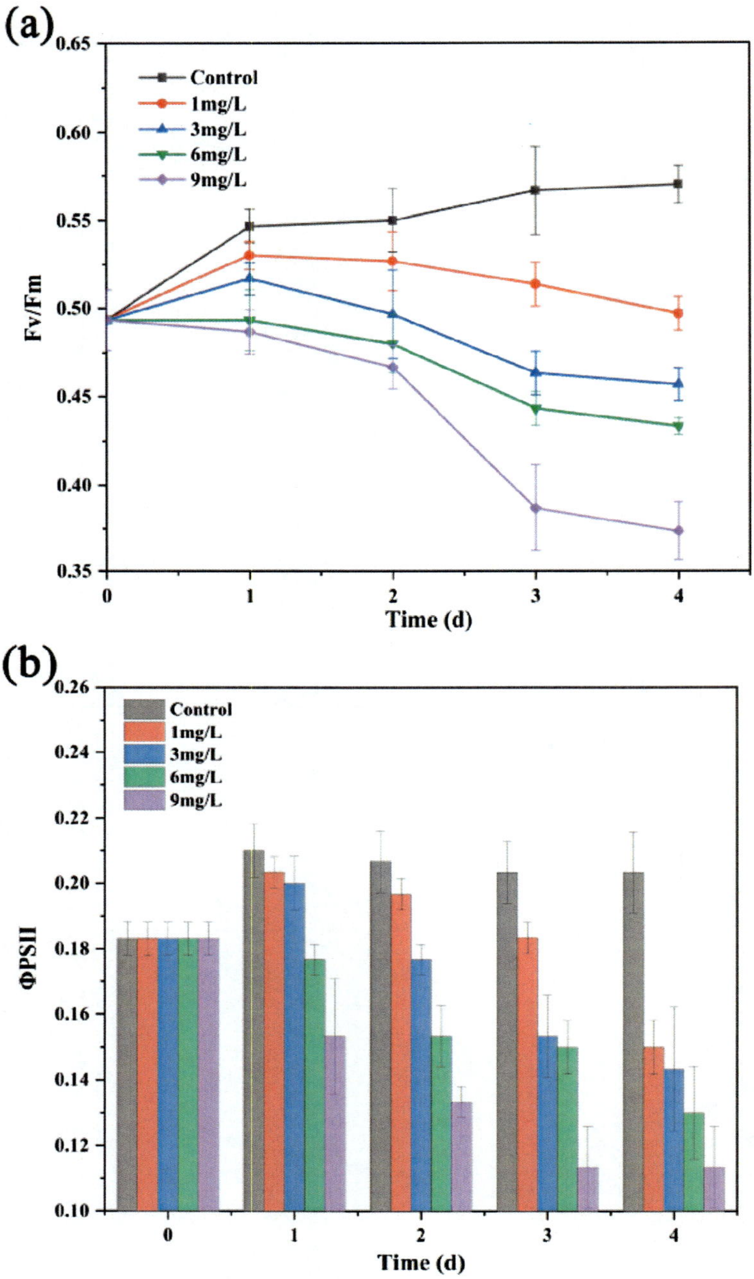

Fig. 3. Effects of Cr (VI) on the Fv/Fm (a) and ΦPSII (b) of *Chlorella vulgaris*

4 Conclusions

Acute toxicity of Cr (VI) to *C. vulgaris* was identified in this work. We found that although *C. vulgaris* can growth at Cr (VI) concentrations of below 6 mg/L while photosynthetic activity decreased. Additionally, Chlorophyll a content responds sharply and decreased in the presence of Cr (VI). Chlorophyll fluorescence analysis implied that the maximum photosynthetic capacity and the actual photochemical efficiency in PSII were inhibited when *C. vulgaris* exposure to Cr (VI), and the growth inhibition enhanced with the increase in concentration and exposure time.

Acknowledgements. This work was supported by Scientific Climbing Plan of Xiamen University of Technology (XPDKT20016).

References

1. Shiller, A.M., Boyle, E.A.: Variability of dissolved trace metals in the mississippi river. Geochim. Cosmochim. Acta **51**(12), 3273–3277 (1987)
2. Kim, S.D., Park, K.S., Gu, M.B.: Toxicity of hexavalent chromium to Daphnia magna: influence of reduction reaction by ferrous iron. J. Hazard. Mater. **93**(2), 155–164 (2002)
3. Rai, V., Vajpayee, P., Singh, S.N., Mehrotra, S.: Effect of chromium accumulation on photosynthetic pigments, oxidative stress defense system, nitrate reduction, proline level and eugenol content of *Ocimum tenuiflorum L.* Plant. Sci. **167**(5), 1159–1169 (2004)
4. Cervantes, C., et al.: Interactions of chromium with microorganisms and plants. Fems. Microbiol. Rev. **25**(3), 335–347 (2001)
5. Jobby, R., Jha, P., Yadav, A.K., Desai N.: Biosorption and biotransformation of hexavalent chromium [Cr (VI)]: a comprehensive review. Chemosphere **207**, 255–266 (2018)
6. Costa, M.: Potential hazards of hexavalent chromate in our drinking water. Toxicol. Appl. Pharm. **188**(1), 1–5 (2003)
7. Browning, C.L., Wise, J.P.: Prolonged exposure to particulate chromate inhibits RAD51 nuclear import mediator proteins. Toxicol. Appl. Pharm. **331**, 101–107 (2017)
8. Ding, T., et al.: Toxicity, degradation and metabolic fate of ibuprofen on freshwater diatom *Navicula sp.* J. Hazard. Mater. **330**, 127–134 (2017)
9. Govindje, E.: Sixty-three years since Kautsky: chlorophyll a fluorescence. Aust. J. Plant. Physiol. **22**(2), 131–160 (1995)
10. Campbell, D., Hurry, V., Clarke, A.K., Gustafsson, P., Oquist, G.: Chlorophyll fluorescence analysis of cyanobacterial photosynthesis and acclimation. Microbiol. Mol. Biol. Rev. **62**(3), 667–683 (1998)
11. Khataee, A., Movafeghi, A., Mojave, R.N., Vafaei, F., Tarrahi, R., Dadpour, M.R.: Toxicity of copper oxide nanoparticles on Spirodela polyrrhiza: assessing physiological parameters. Res. Chem. Intermed. **43**(2), 927–941 (2017)

Analysis of the Deforestation Process of Amazon from During 2001 to 2020 in Peru

Anali Chulluncuy Samaniego(✉), Magaly Godys Avellaneda Ponce, and Steve Dann Camargo Hinostroza

Continental University, San Carlos Avenue 12000, Huancayo, Junin, Peru
72150705@continental.edu.pe

Abstract. Latin America suffers from a major problem of loss of Amazonian Forest cover, thus making Peru one of the ten most diverse countries on the planet where most of our biological diversity is found in forests and these provide us with various ecosystem services and play a very important in nature.

To calculate forest loss, the database of the "Geoforests" platform of the PNCBMCC (National Forest Conservation Program for Climate Change Mitigation) of the Ministry of the Environment was obtained, where Landsat-TM, Landsat- ETM + and Landsat-OLI during the periods 2001–2020.

Based on the evidence, it is concluded that the Peruvian Amazon Forest, over the years, has increased hectares of loss and fragmentation of dangerous forests, predicting that by 2030 there will be almost 4 million hectares lost according to the average of annual growth, being the main causes of agriculture, illegal mining and opening of roads, prevailing the negative aspects such as the loss of biodiversity contributing to climate change.

Keywords: Forest loss · Climate change · Biological diversity

1 Introduction

31% of the planet's surface is forested, which is equivalent to more than 4 billion hectares, during the years 1990 and 2020 there were 420 million hectares lost due to deforestation. Forests bring us many benefits, among them carbon reserves, reduce temperature and help us fight against environmental problems, in addition, the population benefits from forests and we depend on the ecosystem services they provide us [1].

According to SERFOR, worldwide the country with the highest forested area is Peru, occupying the tenth position, in Latin America occupies the second position with the highest forest cover. At the national level, Peru has more than 57% of the area of the humid forest [2].

In 2015, the World Wildlife Fund published "Deforestation in Peru", where they mention that the main origins of deforestation are the construction of lives, changes range in land use such as agriculture, mining commercial, and illegal logging, these problems not only afflict the flora and fauna but also generate greenhouse gases [3].

X. Chen (Ed.): ICESE 2022, ESE, pp. 94–102, 2023.

The illegal acquisition of land by miners, loggers, and farmers, the migration of indigenous populations, the increase in illegal activities, and, above all, the lack of knowledge and cultural values, are some of the main impacts that occur indirectly on the deforestation in humid the forest [4].

In recent years the acceleration of forest loss is very worrying, and the nation's resources are affected, since a great diversity of species of flora and fauna are found in the forests, in addition to offering us goods and services for better quality. of life [5]. As the forests disappear, the ability to have breathable air is lost, as analyzed, the forests are an invaluable source, self-reflection is called for such conservation [6].

MINAM, through the virtual platform GEOBOSQUES, will generate and share monitoring information on changes in the coverage of Peru's forests, this has the purpose of supporting the conservation, prevention, and minimization of deforestation and degradation assigned to the National Program of Forest Conservation for Climate Change Mitigation (PNCBMCC) [7].

2 Methodology

2.1 Place of Study

Peru is located in South America, to the east, it borders Brazil, to the west is the Pacific Ocean, to the north, it borders Ecuador and Colombia, to the southeast, it borders Bolivia.

The study area is limited to the Peruvian tropical humid forests, including the departments of Loreto, Ucayali and Madre de Dios, Amazonas, Cajamarca, Huancavelica, La Libertad, Cusco, San Martín, Huánuco, Pasco, Piura, Puno, Ayacucho and Junín., whose total area reached 68,070,889 ha, with a representation of 64.3% of the country's area as of 2020, the Peruvian territory has a large area of forests, medium annual deforestation, with an inclination that is growing towards rates Increasingly high rates of forest transformation by industrial and small-scale agriculture and artisanal mining [8].

In the Peruvian tropical humid forests, the wet season is between November to May, the dry season is from June to October. During the dry season, the methods for tracking early warnings that use Landsat data show a greater event of forest loss, due to the fact because the presence of clouds and this makes the images more visible [9, 10] (Fig. 1).

2.2 Sampling Method

The loss of forest cover was acquired with quantitative and qualitative data that were acquired from the "Geoforests" platform of the PNCBMCC (National Forest Conservation Program for Climate Change Mitigation) of the Ministry of the Environment, this platform provides information to other different users, through a spatial and temporal analysis [11].

The PNCBMCC provides the reference for annual forest loss between the years 2001–2020 within the framework of the MMCB (Forest Cover Monitoring Module), using Landsat data. Between the period 2001–2011, data from the Landsat-ETM + (Enhanced Thematic Mapper) sensor were used [12], For the 2012–2013 period, Landsat-TM (Thematic Mapper), Landsat-ETM + and Landsat-OLI (Operational Land Imager)

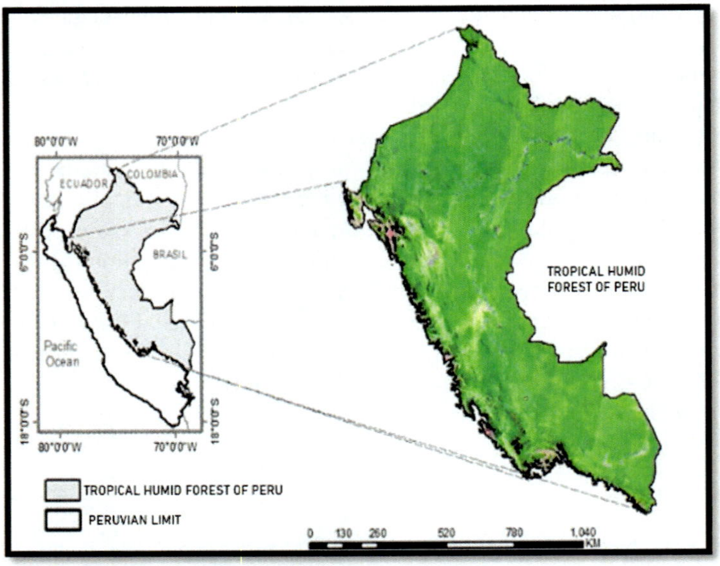

Fig. 1. Study site location **Source**: [9]

data were used [13, 14]. Finally, for the 2014–2020 period, data from Landsat-ETM + and Landsat-OLI were used [15, 16].

In the period 2001–2016, the methodology suggested by the University of Maryland (UMD) is used [8, 12, 13]. This uses a supervised classification algorithm, which means that the user must create training samples manually and based on visual interpretation, this depends on the level of experience of the expert in charge of creating the samples, the experience of each expert cannot be measured and interpretation can be subjective.

In the 2017–2020 period, the Direct Spectral Unmixing (DSU) methodology is used, which is also used in the generation of early warnings of PNCBMCC deforestation. It is an algorithm based on MLME, it assumes that when forest cover is lost the result is a pixel of bare soil, the mixture of soil with dry vegetation or deforestation residues such as logs, which can also be mixed with standing forest, the detection is directly on the image, the expert only decides to detect up to how much percentage of forest cover loss in the pixel. Regarding the annual monitoring process, it is based on a methodological process of decision trees and the results of the loss of forest territory are finally verified and provided feedback by the National Forest and Wildlife Service (SERFOR) [17].

3 Results

Deforestation from the period 2001 to 2020 was 2,636,586.54 ha, with an average annual forest loss of 134,346.94 ha.

According to Fig. 2 shown, the year 2003 was of less forest loss with 72,872 ha, and the year 2020 with 203,272 ha occurred with the greatest loss of very considerable forest concerning the other years.

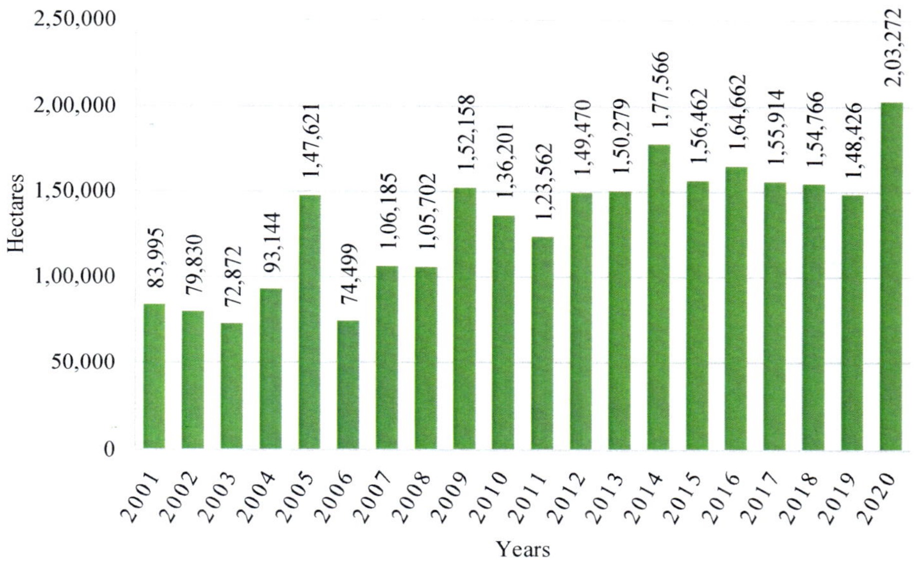

Fig. 2. Annual forest loss 2001–2020

During these 20 years recorded (Fig. 3), the 3 departments with the most forest loss were Loreto (488,198), followed by Ucayali (470,118), and San Martín (467,696), these values being the most representative, unlike the other departments of Peru. Covering the humid forest. The departments with the least loss of humid forest were La Libertad (1,240), Huancavelica (1,276), and Piura (3,384), because these departments cover less territory of Peruvian Amazonian humid forest, unlike the other departments.

Table 1 shows the existing forest area as of 2020 by departments in hectares and percentages, according to the data obtained from the "Geoforests" platform of the PNCBMCC (National Forest Conservation Program for Climate Change Mitigation) of the Ministry of the Environment. It is observed that by the year 2020 the area of the existing Peruvian Amazonian humid forest is 68,070,889 ha, which represents 100% of the Amazonian humid forest of Peru until the year 2020. The departments with the largest area of forest in the year 2020 are Loreto (51.40%), Ucayali (13.59%) followed by Madre de Dios (11.55%). The departments with the smallest area as of 2020 are Huancavelica (0.03%), Piura (0.06%), and La Libertad (0.10%).

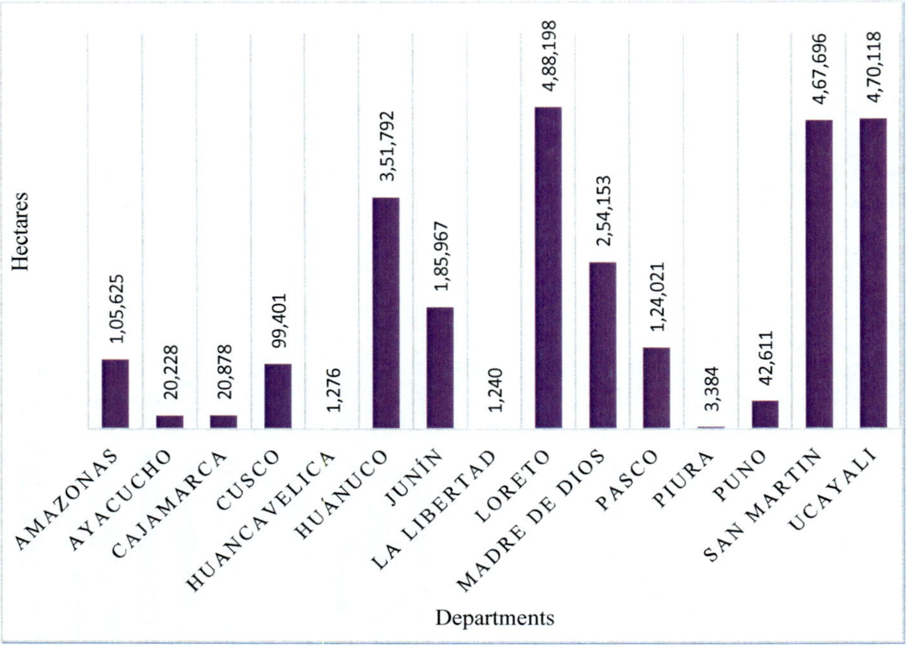

Fig. 3. Forest loss by departments

Table 1. Remaining forest area table as of 2020.

Forest By 2020		
Department	Ha	%
Amazonas	2,814,386	4.13
Ayacucho	205,045	0.30
Cajamarca	344,559	0.51
Cusco	3,047,979	4.48
Huancavelica	17,036	0.03
Huánuco	1,513,104	2.22
Junín	1,816,163	2.67
La Libertad	68,063	0.10
Loreto	34,990,024	51.40
Madre De Dios	7,861,327	11.55
Pasco	1,373,565	2.02
Piura	41,398	0.06
Puno	1,413,756	2.08

(*continued*)

Table 1. (*continued*)

Forest By 2020		
Department	Ha	%
San Martin	3,313,357	4.87
Ucayali	9,251,130	13.59
Total	68,070,889	100.00

Figure 4 shows the dynamics of forest loss growth by 2020, the orange bars are data obtained from the MINAM platform and the red bars show the growth dynamics with an average growth of forest loss of 134,346.94 ha per year. According to the growth rate, the loss of forest by the year 2030 is predicted, estimating that by the year 2030 Peru would have 3,980,055.95 ha of Amazon rainforest lost in Peruvian territory, this being an alarming figure. These data confirm that the growth of forest loss has advanced and will continue to advance if it continues to advance at this rate.

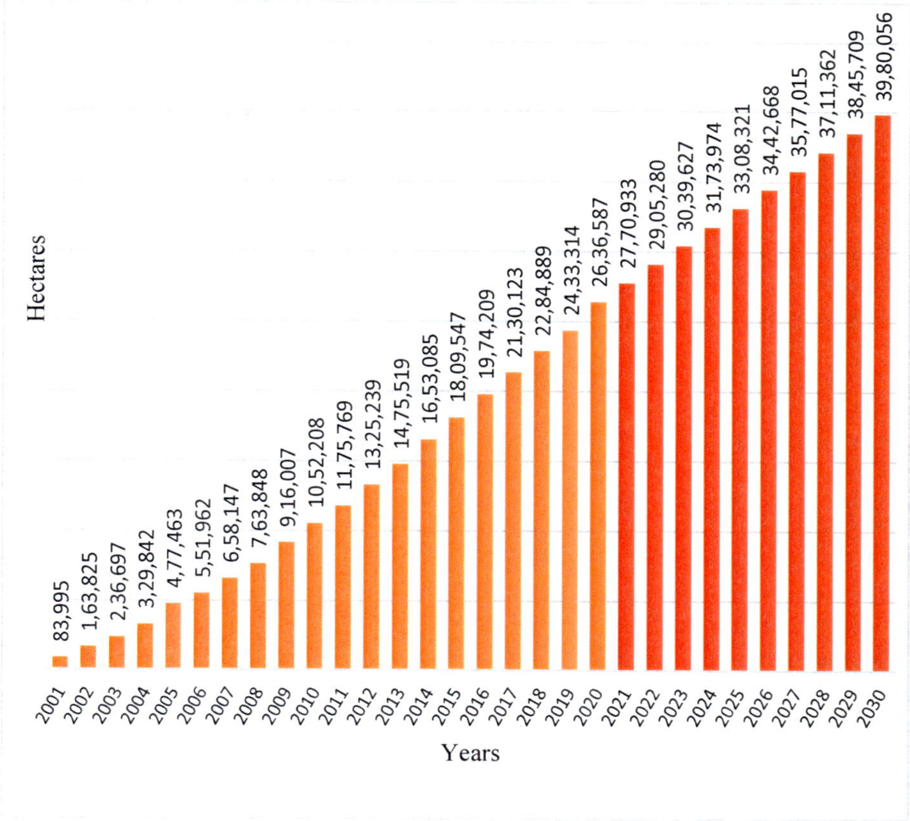

Fig. 4. Increasing forest loss by 2030

4 Discussion

This work demonstrates the main problem in Peru, which is the loss of forest in our territory, also called deforestation, with the departments with the highest loss of the main forest being the departments of Loreto, Ucayali, and San Martin, with alarming figures during these last 20 years and the which if it continues like this will continue to affect the fragmentation of the forest much more as time goes by. The important factors that help the growth of forest loss are due to anthropic causes such as agriculture, agricultural expansion, agricultural migration carried out in an improvised and irrational way by the nearby population, livestock, as well as the openings and construction of access roads and roads [18], These, in turn, disturb the existing flora and fauna in the area, there is also illegal logging for economic purposes provided by forest species and with an end to illegal mining, this being one of the main problems in Madre de Dios [5, 19, 20], It is also caused by forest fires, burning of forests [21, 22].

There is a great discussion of who those who represent and cause deforestation, many reports mention that they are the wood industry, such as chainsaw operators, oil workers, and miners, as well as farmers and ranchers as well as agribusiness, some development projects, central and sectional governments, and road opening, etc. [23, 24].

Faced with the problem of the growth of forest loss, degradation, and/or fragmentation, it is necessary to implement, have environmental commitments, and manage projects for the restoration, management, and conservation of forests and landscapes to avoid negative impacts and factors that limit or hinder the sustainability of this problem, which due to its characteristics hurts the economy and ecology of our country.

5 Conclusions

The analysis shows us that our Peruvian forest is facing severe loss and fragmentation of forests.

According to the data, during these 20 years, Peru suffers from the accelerated growth of loss of hectares of forest. Of the 15 departments where the Peruvian humid forest covers, the main departments where the greatest loss of forest is generated are Loreto with 488,198 ha followed by Ucayali with 470,118 ha, and San Martín with 467,696 ha, these three departments with the highest loss figures during the years 2001–2020.

The year with a high extent of forest loss during the years 2001–2020 was the year 2020 with 203,272 ha, with an average annual forest loss of 134,346.94 ha, representing a 21.35% annual loss growth. Taking into account the average annual loss, the extensive hectares that would be deforested will be inevitable since by the year 2030 there would already be 3,980,056 ha of lost forests, this is an alarming situation, which brings and would bring consequences such as what is being experienced now the climate change.

Although we know that forests provide us with diverse and important ecosystem services and a large part of these territories that provide the highest benefits to our country are being deforested for various reasons. The main causes of forest loss in Peru are caused by anthropic activities, that is, caused by men, such as agricultural expansion and migration, construction and opening of roads, and illegal logging according to previous investigations in deforestation zones.

Due to these problems, it is necessary to promote research and technological improvement in the forestry sector. Even though our country has various normative and regulatory stocks to restore forests, some factors hinder or limit its objective of it. The issue of restoration or mitigation of forest loss should be strengthened by promoting the development of restoration programs and plans at the public or private level.

References

1. FAO: Abridged version of The State of the World's Forests 2022. Forest pathways to green recovery and the creation of inclusive, resilient and sustainable economies, Rome (2022). ISBN 978-92-5-135998-3. https://www.fao.org/3/cb9363es/cb9363es.pdf
2. National Forest and Wildlife Service and National Institute of Statistics and Informatics. Account of Forests of Peru Methodological document, p. 1. Lima (2021). https://www.inei.gob.pe/media/MenuRecursivo/publicaciones_digitales/Est/Lib1811/libro.pdf
3. Smith, J., Schwartz, J.: Deforestation in Peru: how indigenous communities, government agencies, nonprofits and businesses work together to stop forest clearing. WWF Autumn (2015). https://d2ouvy59p0dg6k.cloudfront.net/downloads/la_deforestacion_en_el_peru.pdf
4. SERFOR, MINAGRI, GGGI, DIE and GDI: Interpretation of the dynamics of deforestation in Peru and lessons learned to reduce it. SERFOR, p. 27 (2015)
5. Noblecilla, M.: Effects of forest loss on cricetid rodents in Carpish, Huánuco, Peru. Peruvian J. Biol. 500 (2020). ISSN 1727–9933. http://www.scielo.org.pe/scielo.php?pid=S1727-99332020000400499&script=sci_abstract
6. Leal Pinedo, J.M., Linares Palomino, R.: The Dry Forests of the Northwest Biosphere Reserve (Peru): Tree Diversity and Conservation Status. Caldasia, vol. 27, pp. 195–211 (2005). https://revistas.unal.edu.co/index.php/cal/article/view/39298
7. MINAM: Resolution No. 324–2015 MINAM, Lima (2015)
8. Potapov, P.V., et al.: National satellite-based humid tropical forest change assessment in Peru in support of REDD + implementation, p. 9, 13 (2014). https://doi.org/10.1088/1748-9326/9/12/124012, https://iopscience.iop.org/article/10.1088/1748-9326/9/12/124012
9. Vargas, C., Montalban, J., Leon, A.A.: Early warning tropical forest loss alerts in Peru using Landsat Early warning tropical forest loss alerts in Peru using Landsat, pp. 0–12 (2019). https://iopscience.iop.org/article/10.1088/2515-7620/ab4ec3
10. Hansen, M.C., et al.: Humid tropical forest disturbance alerts using Landsat data Humid tropical forest disturbance alerts using Landsat data (2016)
11. National Forest Conservation Program for Climate Change Mitigation. Forest and Forest Loss. Geoforests (2017). https://geobosques.minam.gob.pe/geobosque/view/index.php
12. MINAM: Protocol of the classification of forest loss cover in Amazon humid forests between the years 2000–2011. Lima, Peru (2015)
13. MINAM: Report of the Loss of the Amazon Rainforests to 2011–2013 (2015)
14. USGS. Landsat 5. https://www.usgs.gov/landsat-missions/landsat-5. Accessed 30 May 2022
15. Ministry of the Environment (MINAM): National Forest Conservation Program for Climate Change Mitigation (PNCBMCC). Monitoring of the loss of Amazon rainforests in 2019. Lima – Peru (2020). http://www.bosques.gob.pe/archivo/Apuntes-del-bosque-3.pdf
16. USGS. Landsat 8. https://www.usgs.gov/landsat-missions/landsat-8. Accessed 30 May 2022
17. MINAM: Cover and loss of Amazon rainforest 2020, pp. 1–19 (2020)
18. Basan Loayza, K.: Multitemporal analysis of the changes of the forest cover and its projection for 2025 in the Villa Carmen Biological Station, Cusco, p. 67. Catholic University Sedes Sapientiae (2020). https://repositorio.ucss.edu.pe/handle/20.500.14095/827?show=full

19. Luque Ramos, L.: Analysis of deforestation in the Peruvian Amazon: Madre de Dios. Innova Education Magazine, vol. 3, p. 209 (2021). https://revistainnovaeducacion.com/index.php/rie/article/download/450/410
20. Alcedo Vega, E., Choque Febres, E.E.: Multitemporal analysis for loss of forest cover 2005 - 2020 by oil palm to minimize Ucayali deforestation, pp. 1–118. Cesar Vallejo University (2021). http://repositorio.ucv.edu.pe/bitstream/handle/20.500.12692/47102/Gutierrez_RS-SD.pdf?sequence=1&isAllowed=y
21. Ministry of Agriculture and Irrigation: Normative Context related to Forest Restoration in Peru (2010). http://repositorio.inia.gob.pe/bitstream/20.500.12955/568/1/Gomez-metodologia_ecosistema.pdf
22. Santamaria Perez, Y.: Deforestation and Its Impact on Biodiversity Loss in the Carpish – HUÁNUCO 2019 Haze Forest, p. 12. HermilioValdizán National University (2021). https://repositorio.unheval.edu.pe/bitstream/handle/20.500.13080/6979/PGA00122S25.pdf?sequence=1&isAllowed=y
23. Burga Ríos, M.: Increase in Deforestation and Its Consequences on the Loss of Biomass in the Forests of the Alto Amazonas Province of the Department of Loreto, 2000–2014, p. 75. Scientific University of Peru (2016). http://repositorio.ucp.edu.pe/handle/UCP/107
24. Zamora Ramírez, M.: Evaluation of the level of deforestation in the concession for the conservation of Angaiza Forests and western sector of the ZoCRE Juninguillo Yanayacu, San Martín, Peru. Catholic University Sedes Sapientiae (2020). https://repositorio.ucss.edu.pe/bitstream/handle/20.500.14095/1247/Zamora_Milagritos_tesis_2020.pdf?sequence=1&isAllowed=y

Evaluation of a Reliable Method for Flash Flood Hazard Mapping in Arid Regions: A Case Study of the Gulf of Suez, Egypt

Mahmoud M. Mansour[1,2,4](✉), Mahmoud Nasr[1,3], Manabu Fujii[4], Chihiro Yoshimura[4], and Mona G. Ibrahim[1,5]

[1] Department of Environmental Engineering, Egypt-Japan University of Science and Technology (E-JUST), Alexandria 21934, Egypt
mahmoud.monir@ejust.edu.eg
[2] Department of Civil Engineering, Faculty of Engineering, Menoufia University, Menoufia 32511, Egypt
[3] Department of Sanitary Engineering, Faculty of Engineering, Alexandria University, Alexandria 21544, Egypt
[4] Department of Civil and Environmental Engineering, School of Environment and Society, Tokyo Institute of Technology, Tokyo, Japan
[5] Department of Environmental Health, High Institute of Public Health, Alexandria University, Alexandria, Egypt

Abstract. Flash floods are severe disasters that devastate the environment, society, and economy in arid regions. Due to the scarcity of observed data from prior incidents, flash flood hazard mapping (FFHM) using Remote Sensing (RS), Geographic Information System (GIS) and hydrological modeling elements has recently become a feasible method for the fundamental management of basins. El-Shamy approach is a straightforward, widely used method for FFHM. This is the first study to investigate the effect of a stream threshold (St) change on the results of El-Shamy approach. Furthermore, runoff peak discharge- and runoff volume-based flash flood hazard degrees (Rp-HD and Rv-HD) were explored. For El-Shamy approach, $St = 0.1\ km^2$ was found to be the best among the considered values for well-distributed estimates of flash flooding potential. A Receiver Operating Characteristic (ROC) curve using historical data was employed to evaluate the considered methods and revealed that El-Shamy suffered from a precision lack, with an Area Under Curve (AUC) equal to 0.36 and 0.45 for low and high hazard basins, respectively, indicating a slight reverse prediction tendency. While Rp-HD and Rv-HD attained a similarly significant and reliable accuracy, with an AUC equal to 1.0 and around 0.865 for low and high hazard basins, respectively. Moreover, Rp-HD was recommended due to its cautious prediction. Accordingly, the FFHM for the study area was presented and associated guiding sustainable management measures were suggested to help decision-makers partially achieve Sustainable Development Goals (SDGs).

Keywords: Flash Flood Prediction · Flooding Map Accuracy · Remote Sensing and Geographic Information System · El-Shamy Approach · Hydrological Modeling

X. Chen (Ed.): ICESE 2022, ESE, pp. 103–117, 2023.

1 Introduction

Flash floods are natural disasters that cause severe environmental, social and economic consequences, including environmental degradation, life losses, cultural heritage threats and infrastructure destruction [1]. Flash floods result from excessive precipitation within a short time and rapid runoff accumulation from upstream mountainous areas [2]. In arid regions, flash floods have recently increased, diminishing development and the ability of the local community to cope with these catastrophes due to a lack of suitable disaster response strategies and unplanned expansion [3]. Moreover, these regions, including the present study area of the Suez Gulf region, suffer from a shortage of monitoring systems and a scarcity of observed data that hamper flash flood forecasts [2]. However, the available tools and data, including Remote Sensing (RS), Geographic Information Systems (GIS), satellite imagery, Digital Elevation Model (DEM), geologic maps, rainfall records and historical data, would facilitate the assessment of flash flood hazard utilizing multiple techniques and, as a result, the development of management strategies. These methods use basic geomorphology indicators such as El-Shamy approach or hydrological modeling outputs such as runoff peak discharge (Rp) and runoff volume (Rv) to estimate and quantify hydrological responses [4].

Many studies employed El-Shamy approach to investigate the spatial potential of flash flooding and recharge potential due to reliance on only three parameters [3–6]. These parameters are susceptible to stream threshold (St) variation and narrowly focus on the description of drainage network and texture descriptions that could cause insufficient or deviated interpretation of the actual conditions. To overcome this issue, Rp and Rv were suggested to comprehensively incorporate relevant morphometry, topography and surface soil factors [2]. Therefore, hydrological modeling is a feasible method to simulate basins' dynamics (Rp and Rv) during any rainfall event. The Hydrologic Engineering Centre-Hydrologic Modeling System (HEC-HMS) was used to predict the rainfall-runoff response in ungauged watersheds, with the Soil Conservation Service Curve Number (SCS-CN) method producing reliable simulations for various basin sizes [4, 7]. Whereas the Receiver Operating Characteristic (ROC) curve was employed for the assessment of the considered methods [8].

Accordingly, the study's main objective was to develop a basis for the mapping of flash flood hazard in the Suez Gulf basins that could be applied to arid regions worldwide and assist planners in implementing sustainable strategies for flash flood management. While the specific targets were the investigation of the effect of stream threshold change on El-Shamy approach's results, the usage of hydrological models to obtain runoff peak discharge- and runoff volume-based flash flood hazard degrees (Rp-HD and Rv-HD), the evaluation of the accuracy of these methods and the utilization of the most reliable method to prepare the map of flash flood hazard.

2 Study Area

The Suez Gulf basin, in the southwestern Sinai Peninsula, Egypt, was chosen for flood hazard investigation because it is a particularly flash flood-prone region and has rarely been studied due to a lack of data. The study area lies between latitudes $27^\circ\ 45'$–$30^\circ\ 03'$

N and longitudes 32◦ 33′–34◦ 12′ E, as shown in Fig. 1a,b. Its area, length and width are all around 13,000 km^2, 275 km and 75 km, respectively. The land levels range from 0 m on the Suez Gulf shore (southwest) to 2625 m in the mountainous region (east). Almost the whole study area is bare and desert land. The average annual rainfall was about 20 mm [9].

3 Methodology

Flash flood hazard mapping (FFHM) relied on two techniques. First, a few morphometric parameters were calculated to define flash flood potentials with El-Shamy approach. Second, hydrological modeling was used to estimate the Rp and Rv along with the hydrograph of rainfall-runoff simulation for each basin. Therefore, Rp and Rv were utilized to assess the flash flood hazard degree (FFHD) based on equal interval classification for these parameters. The current study utilized different spatial datasets for the Suez Gulf region, including satellite imagery, DEM, soil data, rainfall and historical data for flash floods in the study area.

3.1 Data Acquisition and Processing

The Shuttle Radar Topography Mission (SRTM) and Enhanced Thematic Mapper Plus (ETM +) on Landsat-7 freely provided the required satellite imagery and DEM with a 30 m resolution [10]. These data were downloaded from (https://earthexplorer.usgs.gov/), Fig. 1a,b. The DEM was a pivotal input for ArcGIS version 10.4.1 and HEC-HMS version 4.8 to define basin boundaries, delineate drainage networks and derive hydrographs, Fig. 1b,c. The study area contains 56 basins with an area larger than 18 km^2, such as El-Raha, Sudr, Wardan Gharandel, Baebae, Sedery, Feran and El-Aawag basins. Satellite imagery played a crucial role in checking the alignment of the drainage course and recognizing land use/land cover (LULC) in the studied region. Surfaces Soil distribution was collected from the Egypt soil map presented by Eliwa et al. [1], Fig. 1d. Infiltration and abstraction losses during rainfall events were approximated depending on soil type and LULC distribution. The rainfall pattern, on the other hand, could be retrieved from the metrological data of nearby stations, provided by Dadamouny and Schnittler [9]. It was essential for the hydrological models to determine the cumulative rainfall depth of the design storm. Over 41 years, till 2020, flash flood events in the study area were compiled from available newspapers, reports, dissertations, online websites, and published articles [1, 3]. FFHD for the Suez Gulf basins was identified relying on the frequency and magnitude of these data, as shown in Table 1 [11, 12]. It was employed for subsequent evaluation of the considered methods: El-Shamy approach, Rp-HD and Rv-HD.

Table 1. FFHD classification based on the reported past flash flood events.

FFHD	Very High	High	moderate	Low	Very low
Basins	30	11, 19, 23 and 35	14, 27 and 28	21, 22 and 24	9, 10, 12, 13, 15, 20, 36 and 37

3.2 El-Shamy Approach

This approach was first developed for flash flooding and groundwater recharge potential estimation by El-Shamy [13]. It is a simple method as it relies on only 3 morphometric parameters: mean bifurcation ratio (Rb), stream frequency (F) and drainage density (D) [14]. Table 2 shows the calculation formulas for these parameters. This method includes two relationships (Rb vs D and Rb vs F). Applying this relationship separately to each basin elucidates the classification of both flash flood hazard and groundwater recharge potential. Each relationship chart has two curves that divide the domain into three fields, as subsequently illustrated. These fields, referred to as A, C and B, reflect the low, moderate, and high flash flooding potentials, respectively, and vice versa for groundwater recharge conditions. Due to the limited number of parameters (only 3) in this approach, it could be considered a poor classifier. Moreover, the values of the involved parameters and, as a result, their outcomes could be highly influenced by the change in stream threshold (St), which is the minimum required drainage area to originate first streams. St is a user-defined input that directly affects the drainage network configurations, namely, stream lengths and numbers, and initial point locations and numbers. Thus, 4 stream thresholds (Sts) were selected, including 0.3, 0.2, 0.1 and 0.05 km^2 to explore the effect of St change on the performance of El-Shamy approach and find the better value for FFHM.

Table 2. The methods and equations of estimation for the main characteristics of drainage network and texture.

No	Morphometric parameter	Formula	Unit	References
1	Basin area (A)	GIS software analysis	km^2	[15]
2	Drainage density (D)	$D = \sum L_u/A$; where L_u = stream length of order 'u'	km/km^2	[16]
3	Stream frequency (F)	$F = \sum N_u/A$; where N_u = number of streams of order 'u'	$1/km^2$	[16]
4	Mean bifurcation ratio (Rb)	$Rb = \sum (N_u/N_{u+1})/(U-1)$; where N_{u+1} = number of streams of next higher order 'u + 1' and U = basin's maximum stream order	Dimensionless	[17]

3.3 Hydrological Modeling

Hydrological modeling is a practical means of quantifying the hydrological response of basins. So, HEC-HMS version 4.8 was employed for the rainfall-runoff simulation of watershed systems in the whole study area. The US Army Corps of Engineers created the HEC-HMS, which is a semi-distributed and physically based model [18]. This

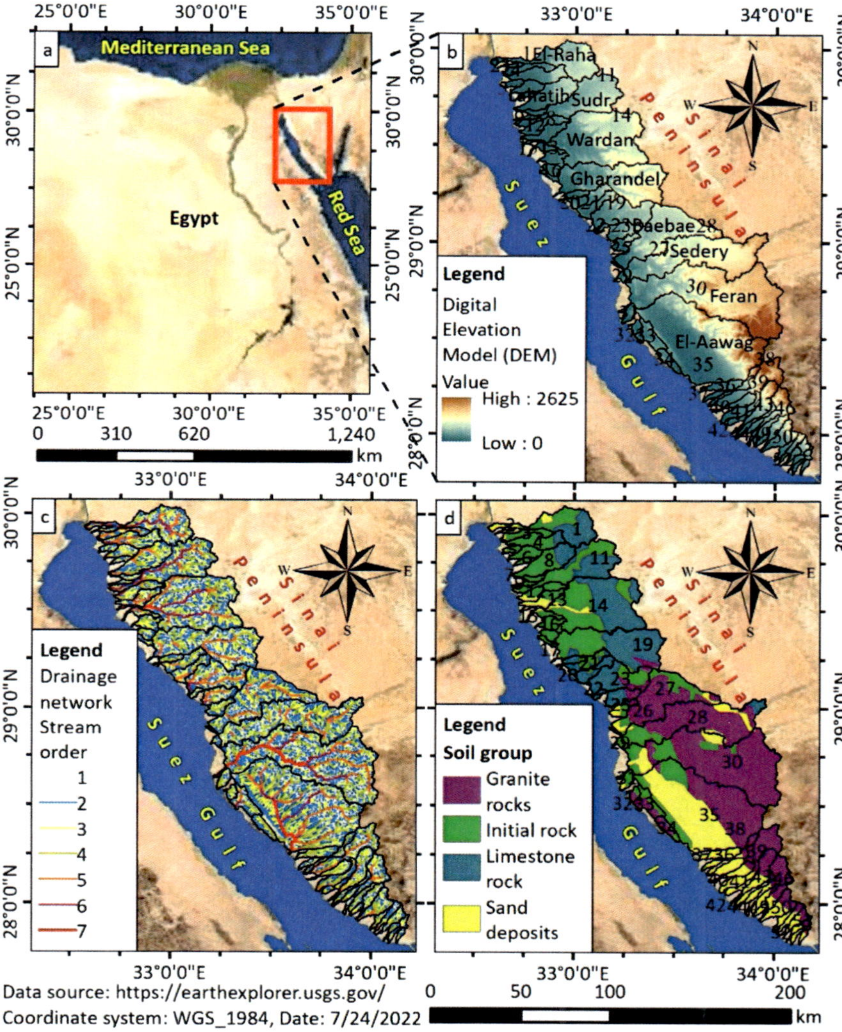

Fig. 1. Multiple maps for the Suez Gulf basins: a) satellite imagery and location, b) DEM and basin boundaries, c) drainage networks and stream orders for St 0.1 km^2 and d) surface soil.

model was developed to be well adapted to a variety of geographic locations and cope with a wide range of challenges, such as small, large and ungauged basins [4]. The simulations facilitate water resources management, flash flood hazard assessment and future developments [7]. A basin model, time-series data, meteorological model and control specifications are the main sub-models of the HEC-HMS software. It allows for employing several infiltration, unit hydrograph and flood routing methods [18]. The Curve Number (CN) method of the Soil Conservation Service (SCS) was utilized as a loss model. The CN is generated based on LULC classes and hydrological soil groups.

Therefore, runoff depth could be computed using Eqs. (1–3):

$$Q = \frac{(P - I_a)^2}{P - I_a + S} \, for \; P > I_a; \; else \; Q = 0 \tag{1}$$

$$S = 25.4\left(\frac{1000}{CN} - 10\right) \tag{2}$$

$$I_a = 0.2S \tag{3}$$

where Q is the excess rainfall or surface runoff (mm), P is the storm total precipitation (mm), S is the maximum potential retention or surface storage (mm) and I_a is the initial abstraction and infiltration before runoff (mm).

The unit hydrograph of SCS was chosen as the transform method in the HEC-HMS to convert rainfall to streamflow hydrograph, which required the lag time parameter in minutes. It is the duration between the mass center of surplus precipitation and the peak of the hydrograph, which is equal to 0.6 of the time of concentration, and can be directly calculated for each watershed by Eq. (4), described by the Natural Resources Conservation Service (NRCS) [19]:

$$L = \frac{l^{0.8}(S + 1)^{0.7}}{1900Y^{0.5}} \tag{4}$$

where L is the lag time (h), l is the flow length (ft), Y is the average land slope of the watershed (%) and S is in inches.

The Muskingum model was used as the routing method in these ungauged basins, which necessitated a travel time of wave (K) that reached 3.36 h and a discharge weight coefficient (X) varying from 0 to 0.5 [20]. Between 1994 and 2014, the higher extreme daily precipitations for the study area ranged from below 65 mm to above 85 mm with limited spatial variation [9]. Furthermore, a design storm of cumulative values of 75 mm/day was recommended for 50- and 100-year return periods with significant probability by Abdel-Lattif and Sherief [21]. So, a rainfall depth of 75 mm/day was assigned to the hydrologic models following the SCS Type II design storm hyetograph [18].

3.4 Evaluation

A Receiver Operating Characteristic (ROC) curve was employed to quantitatively assess the considered methods' precision, which was rarely utilized for similar FFHD assessment methods in the explored articles. ROC curve is an accepted measure to illustrate prediction performance and compare different models by offering a visual depiction of the equilibrium of the true and false positive rates (sensitivity vs. 1-specificity) for every potential suitability value [8]. The curve is a plot in two dimensions, where the false positive rate is on the X-axis and the true positive rate is on the Y-axis, as subsequently shown. They were calculated using Eqs. (5–6):

$$\text{True positive rate(Sensitivity)} = \frac{TP}{TP + FN} \tag{5}$$

$$\text{False positive rate}(1 - \text{Specificity}) = \frac{FP}{TP + FN} \quad (6)$$

where TP is true positive, FP is false positive and FN is false negative.

The ROC curve quantifies the performance, accuracy or overall quality of the models using the area under the curve (AUC). It ranges from 0 to 1.0. Higher model accuracy is associated with an increased value of AUC. The model's accuracy is considered significant and reliable for an AUC value approaching 1 and vice versa for an AUC value nearing 0.5. To produce the ROC curves and evaluate the quantitative accuracy of the studied methods, the historical data based-FFHD was used.

4 Results and Discussions

Box plots are a convenient way to visually examine one or more sets of data while using less space to compare data distributions. As seen in Fig. 2, it provides information about the data's structure, range, median, and mean, as well as outliers (extreme values). Figure 2 depicts the impact of changing St on each basin's total stream lengths and number, and maximum stream order. Figure 2a,b demonstrate that the distribution forms of total stream length and number records were right-skewed for all Sts due to the increased number of smaller basins, while Fig. 2c illustrates that the distributions were almost symmetric for the basin's maximum stream order. Total stream lengths and numbers witnessed exponential increases as St decreased from 0.3 km^2 to 0.05 km^2, confirming the findings of Reddy et al. [22], while the basin's maximum stream order was raised slightly. Decreasing St resulted in more and/or closer to basin boundary stream initial points, resulting in longer streams with the first order, new stream generation and/or relatively limited order upgrade of some existing streams.

Figure 3 depicts the two relationships of El-Shamy approach (Rb Vs D and Rb vs F charts) for flash flooding potential estimation as previously mentioned and the distribution of the studied basins among the three flash flood potentials for various Sts. Figure 3a shows an incremental increase in D values as St reduces owing to the increase in total stream lengths with constant basin areas, whereas Fig. 3b illustrates

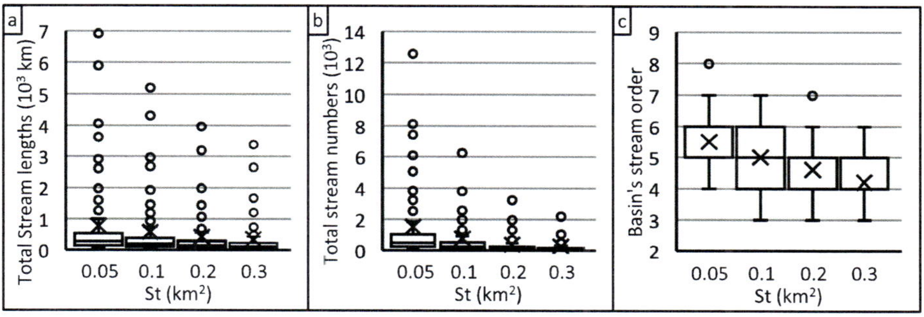

Fig. 2. Box plot for the effect of St change on some basin characteristics, a) total stream lengths, b) total streams number and c) maximum stream order.

an exponential rise in F values for lowering St due to the considerable increase in total stream numbers, meeting the outcomes of Reddy et al. [22]. The Rb changed slightly, with a slight increase in lower values and an obvious decrease in higher values, leading to a narrower range of records due to St lowering. The mean bifurcation ratio (Rb) was affected by the bifurcation ratio of the two highest-order streams, which tended to be either extremely low or extremely high. So, decreasing St often increased the number of stream orders and then lowered the weight and impact of these extreme values.

For the Rb vs D chart, the basins' distribution over the three zones shifted by a step forward from St of 0.3 to 0.05 km^2. The counts of basins in zones A and C for St of 0.3 km^2 were nearly identical to those in zones C and B for St of 0.05 km^2, respectively. The only basin in zone B for St of 0.3 km^2 also had a corresponding basin in zone A for St of 0.05 km^2, indicating a distribution gap. For Sts of 0.2 and 0.1 km^2, the distributions of basins over the three hazard degrees were better than those of Sts of 0.3 and 0.05 km^2, but the distribution of St of 0.1 km^2 was considered the best as the differences in basin count across all zones were relatively low. The basins' distribution over the three zones in the Rb vs F chart depicted a worse gap and shortage with no basin lying in zone B for the first three Sts of 0.3, 0.2, and 0.1 km^2 and no basin lying in zone A for the St of 0.05 km^2. For St of 0.3 km^2, the number of basins in zone A was slightly more than that of zone C. It gradually changed till the basin count in zone C reached 2.5 times the basin count in zone A for St of 0.1 km^2. Then, the number of basins in zone C was nearly 1.5 times that of zone B for St of 0.05 km^2.

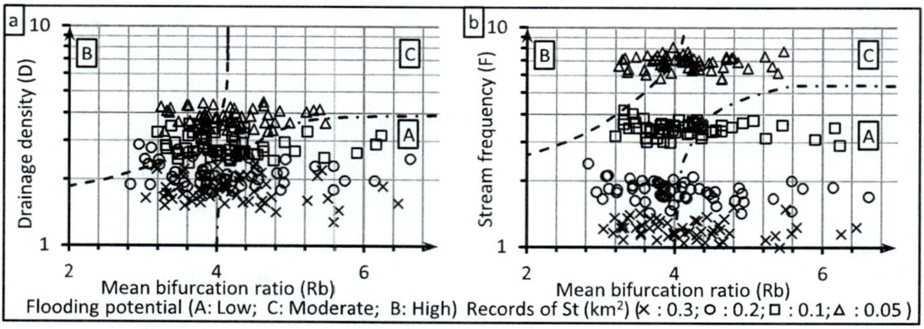

Fig. 3. Estimation of the flash flooding potential of the Suez Gulf basins for different Sts using the two relationships of El-Shamy approach: a) Rb vs D and b) Rb vs F.

The higher hazard degree for each basin of the two charts was considered for the overall assessment. Figure 4 visualizes the allocation of basin count and basins area percentage (%) to the 3 flash flooding potentials according to El-Shamy approach for the 4 Sts as histograms. In terms of basin count and area percentage, the results of St of 0.3 km^2 contradicted that of St of 0.05 km^2 for low and high hazard degree classes. For Sts of 0.2 and 0.1 km^2, the allocation followed better distribution. The classification of St of 0.1 km^2 was selected for subsequent evaluation because it had the closest pattern to the normal distribution regarding basin count and basins area percentage. Accordingly, a change in stream threshold could adversely manipulate the result of this approach. Furthermore, neither the approach nor the previous studies provided any assumptions about the selection of an appropriate stream threshold value, which represented an obvious shortage. However, the performed procedures in the present study by examining many stream thresholds could be a practical option for this issue.

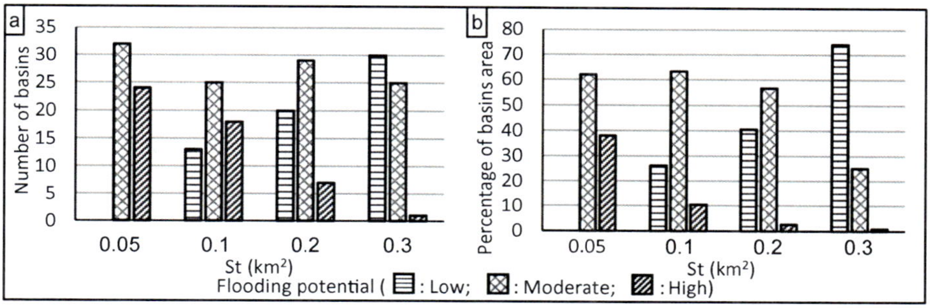

Fig. 4. Distribution of the basins over the three flash flooding potentials according to El-Shamy approach for the 4 Sts in the form of a) number of basins and b) percentage of basins area.

Using the HEC-HMS model resulted in the flash flood hydrographs for the Suez Gulf basins and the 9 flash flood hydrographs for the largest basins were plotted in Fig. 5. The runoff peak discharges and runoff volumes were identified at the outlet of each basin by the interpretation of these graphs. The peak discharges and volumes of runoff illustrate the basin's capabilities for runoff generation from rainfall and are always employed in the design of hydraulic structures. Figure 5 shows that Feran basin produced the highest peak flow (1648.5 m^3/s) and runoff volume (65.7 Mm^3), exceeding El-Aawag basin (921.9 m^3/s and 47.27 Mm^3) with a larger area and flattened peak discharges because the soil of El-Aawag basin was characterized by higher infiltration capacity than that of Feran basin. The peak discharge of Sedery (951.9 m^3/s) basin was slightly higher and considerably later than that of Baebae basin (892.06 m^3/s) because of the more elongated shape and larger basin area of Sedery basin. The hydrograph features, viz., runoff peak discharge (Rp) and runoff volume (Rv), and associated flash flood hazard degree are summarized in Table 3 for all basins.

Peak discharge of runoff varied from 6.9 m^3/s for basin 37 to 1648.5 m^3/s for basin 30 and runoff volume ranged from 0.09 Mm^3 for basin 10 to 65.7 Mm^3 for basin 30, as described in Table 3. The Rp and Rp-HD of basin 39 in the southern zone were comparatively greater than those for basins with similar or larger basin areas and Rv, such as basins 1, 8, 11 and 23 in the northern area. This figure was ascribed to the soil's lower infiltration capacity, higher topography, higher land surface slopes, and more circular shape of the main catchment for basin 39 than for these northern basins. Table 3 reveals that the runoff peak discharge-based flash flood hazard degree (Rp-HD) was always equal to that of the runoff volume-based flash flood hazard degree (Rv-HD), except for 9 basins. Rp-HD was higher than Rv-HD by 1 step for 8 basins and 2 steps for 1 basin. This trend implied that Rp provided a more cautious prediction than Rv. Moreover, Rps are considered the main trigger and cause of catastrophic repercussions of flash flood events owing to the limited time available for preparedness before the occurrence of Rp.

Table 3. The estimated runoff peak discharge (Rp) and runoff volume (Rv) with the corresponding flash flood hazard degrees (Rp-HD and Rv-HD) and basin area.

No.	Area (Km^2)	Rp (m^3/s)	Rv (Mm^3)	Rp-HD	Rv-HD	No.	Area (Km^2)	Rp (m^3/s)	Rv (Mm^3)	Rp-HD	Rv-HD
1	469.0	345.42	11.40	2	1	30	1780.9	1648.51	65.70	5	5
2	21.4	14.23	0.39	1	1	31	48.2	41.87	1.06	1	1
3	81.8	85.50	2.06	1	1	32	32.7	90.93	1.26	1	1
4	181.6	186.74	4.74	1	1	33	123.0	229.21	4.01	1	1
5	31.3	33.21	0.76	1	1	34	66.6	142.58	2.33	1	1
6	79.6	83.16	2.10	1	1	35	1906.6	921.90	47.27	4	4
7	26.4	33.82	0.70	1	1	36	85.1	20.83	0.89	1	1
8	281.1	292.67	7.41	2	1	37	30.4	6.94	0.32	1	1
9	35.0	43.25	0.92	1	1	38	76.8	103.59	1.96	1	1
10	20.8	23.96	0.09	1	1	39	279.8	555.18	10.53	3	1
11	565.2	430.99	14.82	2	2	40	72.9	54.90	1.28	1	1
12	113.7	113.71	2.92	1	1	41	99.9	47.93	1.60	1	1
13	109.0	76.21	2.08	1	1	42	36.8	12.57	0.39	1	1
14	1156.9	787.51	29.03	4	3	43	157.8	268.48	4.73	2	1
15	50.5	47.28	1.06	1	1	44	49.5	47.93	1.00	1	1
16	80.4	104.06	2.12	1	1	45	33.9	13.67	0.36	1	1
17	44.9	69.30	1.18	1	1	46	169.7	354.57	6.41	2	1
18	24.4	44.59	0.64	1	1	47	56.5	33.27	0.82	1	1
19	866.9	672.67	23.26	3	2	48	19.1	7.40	0.20	1	1
20	42.5	62.81	1.12	1	1	49	73.5	99.28	1.77	1	1
21	117.0	126.75	3.09	1	1	50	127.2	176.56	3.08	1	1
22	107.4	157.03	2.83	1	1	51	33.3	18.35	0.42	1	1
23	356.4	409.84	10.12	2	1	52	49.9	50.43	0.96	1	1
24	18.7	29.53	0.49	1	1	53	63.4	76.02	1.43	1	1
25	121.3	189.31	3.68	1	1	54	22.0	8.76	0.26	1	1
26	75.9	107.27	1.76	1	1	55	39.2	57.54	1.03	1	1
27	721.5	892.06	25.68	4	3	56	17.9	11.69	0.28	1	1
28	1068.2	951.89	36.78	4	4	Min	17.9	6.9	0.09	1	1
29	20.2	37.67	0.53	1	1	Max	1906.6	1648.5	65.7	5	5

Figure 6 illustrates the ROC curve and AUC for the three prediction methods (e.g., Rv-HD, Rp-HD and El-Shamy approach) in classifying the basins with low and high FFHDs. Figure 6a shows that Rv-HD and Rp-HD provided optimum accuracy in defining the basins with low FFHD with an AUC equal to 1.00, but El-Shamy approach had a poor performance with an AUC near 0.5 (0.36), indicating the tendency to detect more basins with higher FFHDs than basins with low FFHD as basins with low FFHD. Figure 6b shows that Rv-HD and Rp-HD could reliably and consistently predict basins with high FFHD with an AUC near 1.00 (around 0.865), implying remarkable and outstanding performance. On the other hand, El-Shamy showed almost random classification with an AUC equal to 0.45, suggesting unreliable and misleading results and agreeing with the conclusions of Nasir et al. [6]. Due to the almost identical accuracy of Rv-HD and Rp-HD and the discreet classification of Rp-HD, Rp-HD was considered the most significant and reliable method and it was recommended to present the FFHD map, as shown in Fig. 7.

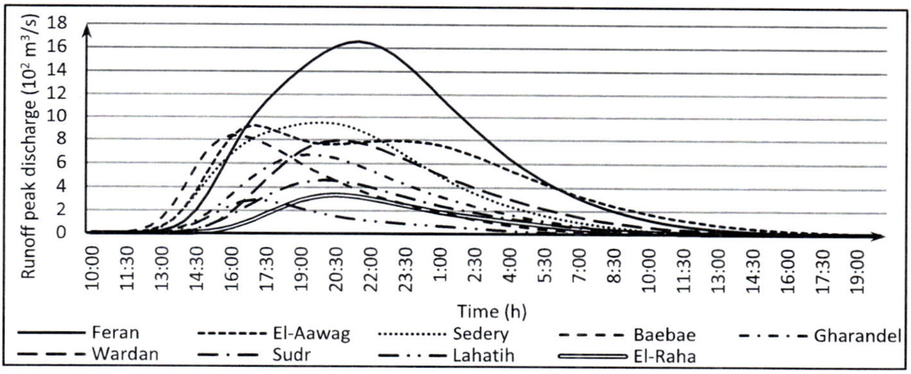

Fig. 5. Flash flood hydrographs for the 9 largest basins in the study area.

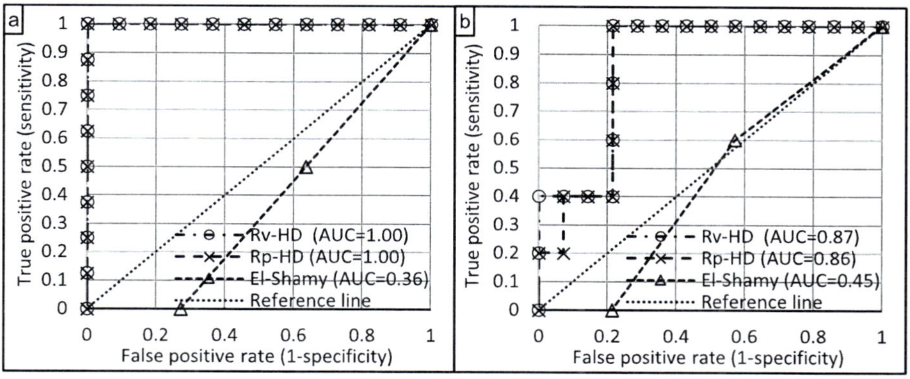

Fig. 6. The receiver operating characteristic (ROC) curve and area under the curve (AUC) based on the assessment of historical data for Rv-HD, Rp-HD and El-Shamy approach regarding the prediction of basins with a) low FFHD and b) high FFHD.

Figure 7 shows the FFHM based on runoff peak discharges for all basins in the study area. Rp considered the incorporation of many parameters, involving catchment area, basin shape, land surface slope, basin relief, mainstream properties and surface soil characteristics. The flash flood-prone basins were relatively classified, hazard-wise, into five categories, namely, very high, high, moderate, low and very low. This map could facilitate suggestions for mitigation strategies to reduce the waste of fresh water and the devastating impacts caused by flash floods. Furthermore, it would aid in the retarding of saltwater intrusion in this coastal region by allowing for local infiltration, replenishment of groundwater storage, and a rise in groundwater levels without the need for costly vertical physical barriers [23–28].

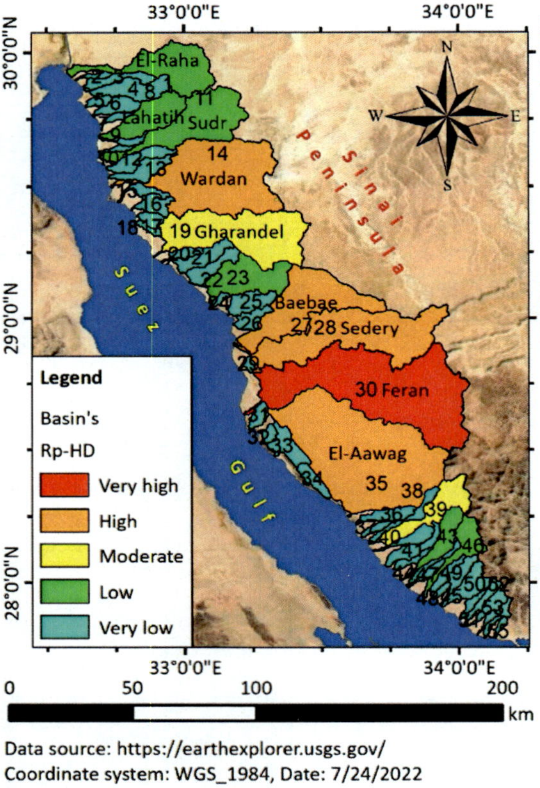

Fig. 7. Flash flood hazard degree map based on Rp-HD for the whole study area.

Several measures could be implemented in Feran, El-Aawag, Sedery, Baebae and Wardan basins with higher FFHD in order to alleviate the destructive consequences of flash flood episodes and store fresh water for municipal, agricultural and economic purposes. In arid areas having a hilly and rugged topography, cisterns, diversion dams and storage reservoirs could be utilized to harvest fresh water for household consumption. The construction of dikes would safeguard the cultivated areas along its downstream. Furthermore, special cisterns should be constructed in the flash flood-affected zone to store runoff and collect the subsurface seepage of the upstream zones. Then, the geological characteristics of the downstream areas of El-Aawag and Wardan basins are regarded as appropriate locations for the artificial recharge of groundwater where the alluvial layer is thick, permeable and low in topographic gradient, forming an unconfined aquifer with high infiltration conditions. Eventually, the following appropriate measures were proposed for decision-makers to develop a suitable management strategy considering the three pillars of sustainable development, particularly, the environment, society and economy:

- construction of detention dams at the upper segments of main streams.
- excavation of artificial lakes near the outlet of the catchment to store runoff in the shallow aquifer of the main valley.
- building of sediment traps along the hazardous basins.
- insertion of culverts at the intersection of a flash flood route with vital highways and pipelines.
- application of land use planning policy to direct development away from the high hazard area.
- activation of early warning systems for flash flood-prone zones.

The proposed management measures provide significant benefits to multiple targets of some Sustainable Development Goals (SDGs), including SGD 6 "Clean water", SDG 8 "Economic growth", SDG 11 "Sustainable cities", SDG 14 "Life below water", and SDG 15 "Life on land".

5 Conclusion

The drainage network and basin boundaries were delineated using SRTM-DEM with a 30 m resolution and correctly traced on enhanced satellite imagery using ArcGIS. 56 basins in the Suez Gulf region were identified, such as Sudr, Gharandel, Sedery, Feran and El-Aawag. The main parameters of the drainage network, utilized in El-Shamy approach, were computed for four different Sts, viz., 0.3, 0.2, 0.1 and 0.05 km^2. These morphometric parameters included mean bifurcation ratio (Rb), stream frequency (F) and drainage density (D). The decrease of stream thresholds led to a significant increase in these parameters' values, except for the mean bifurcation ratio, which just witnessed a slight decrease in its values range. The effect of stream threshold change on El-Shamy approach's results was investigated and the best value (0.1 km^2) was suggested for flash flood potential classification. Hydrological models for all basins were built in HEC-HMS using DEM, soil distribution, LULC and a synthetic storm to obtain flash flood hydrographs along with Rp and Rv for FFHD classification. The basins were classified into five FFHDs based on the Rp and Rv values separately. The ROC was employed based on historical data to evaluate the accuracy of the proposed methods for FFHD prediction. Rv-HD and Rp-HD provided almost similar reliable precision with an AUC ranging from around 0.865 to 1.0 and were significantly better than El-Shamy approach with an AUC ranging from 0.36 to 0.45, indicating accuracy lack. Moreover, Rp-HD was considered a discreet classifier. So, the FFHD map based on Rp-HD was presented and appropriate management measures were suggested for decision-makers, achieving many targets of the SDGs.

Acknowledgements. The first author is very grateful to the Egyptian Ministry of Higher Education (MoHE) for providing financial support in the form of a Ph.D. scholarship. Also, thanks to the Japan International Cooperation Agency (JICA) for providing all the facilities, and equipment to accomplish this research.

References

1. Eliwa, H.A., Murata, M., Ozawa, H.: Post Aswan High Dam flash floods in Egypt: Causes, consequences and mitigation strategies. Bull. Center Collaborat. Community Naruto Univ. Educ. **29**(2), 173–186 (2015)
2. Helmi, A.M., Zohny, O.: Flash flood risk assessment in Egypt. In: Negm, A.M. (ed.) Flash floods in Egypt. ASTI, pp. 253–312. Springer, Cham (2020). https://doi.org/10.1007/978-3-030-29635-3_13
3. Youssef, A.M., Pradhan, B., Hassan, A.M.: Flash flood risk estimation along the St Katherine road, southern Sinai, Egypt using GIS based morphometry and satellite imagery. Environ. Earth Sci. **62**(3), 611–623 (2011)
4. Abdelkader, M.M., Al-Amoud, A.I., El Alfy, M., El-Feky, A., Saber, M.: Assessment of flash flood hazard based on morphometric aspects and rainfall-runoff modeling in Wadi Nisah, central Saudi Arabia. Remote Sens. Appl. Soc. Environ. **23**(100562) (2021)
5. Mansour, M.M., Ibrahim, M.G., Fujii, M., Nasr, M.: Recent applications of flash flood hazard assessment techniques: case studies from Egypt and Saudi Arabia. Adv. Eng. Forum **47**, 101–110 (2022). https://doi.org/10.4028/p-03z404
6. Nasir, M.J., Iqbal, J., Ahmad, W.: Flash flood risk modeling of swat river sub-watershed: a comparative analysis of morphometric ranking approach and El-Shamy approach. Arab. J. Geosci. **13**(20), 1–19 (2020). https://doi.org/10.1007/s12517-020-06064-5
7. Ben Khélifa, W., Mosbahi, M.: Modeling of rainfall-runoff process using HEC-HMS model for an urban ungauged watershed in Tunisia. Modeling Earth Syst. Environ. 1 (2021). https://doi.org/10.1007/s40808-021-01177-6
8. Lin, K., et al.: Assessment of flash flood risk based on improved analytic hierarchy process method and integrated maximum likelihood clustering algorithm. Journal of Hydrology, 584(124696), (2020)
9. Dadamouny, M.A., Schnittler, M.: Trends of climate with rapid change in Sinai Egypt. J. Water Clim. Change **7**(2), 393–414 (2016)
10. Mansour, M.M., Nasr, M., Fujii, M., Yoshimura, C., Ibrahim, M.G.: Quantification of flash flood runoff volume using morphometric parameters towards sustainability. In: Ujikawa, K., Ishiwatari, M., Hullebusch, E.V. (eds.) Environment and Sustainable Development. ACESD 2022. Environmental Science and Engineering. Springer, Singapore, (Forthcoming 2023)
11. Mansour, M.M., Ibrahim, M.G., Fujii, M., Nasr, M.: Sustainable development goals (SDGs) associated with flash flood hazard mapping and management measures through morphometric evaluation. Geocarto International (2022). https://doi.org/10.1080/10106049.2022.2046868
12. EXCIMAP, European exchange circle on flood mapping: Handbook on good practices for flood mapping in Europe. European Environment Agency, Copenhagen, Denmark (2007)
13. El-Shamy, I.: New approach for hydrological assessment of hydrographic basins of recent recharge and flooding possibilities. In: 10th Symposium Quaternary and Development proceedings on proceedings, p. 15, Mansoura, Egypt, April 1992
14. Mansour, M.M., Nasr, M., Fujii, M., Yoshimura, C., Ibrahim, M.G.: Identification of a practical method and a set of morphometric parameters for flash flood potential prioritization. In: 14th International Conference on Hydroscience & Engineering (ICHE2022) on proceedings, pp. 603–615. Izmir, Turkey (2022)
15. Schumm, S.A.: Evolution of drainage systems and slopes in badlands at Perth Amboy, New Jersey. Bull. Geol. Soc. Am. **67**(5), 597–646 (1956)

16. Horton, R.E.: Erosional development of streams and their drainage basins; hydrophysical approach to quantitative morphology. Geol. Soc. Am. Bull. **56**(3), 275–370 (1945)
17. Strahler, A.N.: Revision of Hortons' quantitative factors in erosional terrain. Trans. Am. Geophys. Union **34**, 345–365 (1953)
18. Scharffenberg, W.A.: Hydrologic modeling system HEC-HMS: User's manual version 4.2. U.S. Dept. of Agriculture, Hydrologic Engineering Center, Davis, CA, USA (2016)
19. Mockus, V.: Watershed lag. U.S. Dept. of Agriculture, Soil Conservation Service, Washington, DC (1961)
20. Song, X., Kong, F., Zhu, Z.: Application of Muskingum routing method with variable parameters in ungauged basin. Water Sci. Eng. **4**(1), 1–12 (2011)
21. Abdel-Lattif, A., Sherief, Y.: Morphometric analysis and flash floods of Wadi Sudr and Wadi Wardan, Gulf of Suez, Egypt: using digital elevation model. Arabian J. Geosci. **5**, 181–195 (2012)
22. Reddy, G.P.O., Kumar, N., Sahu, N., Singh, S.K.: Evaluation of automatic drainage extraction thresholds using ASTER GDEM and Cartosat-1 DEM: a case study from basaltic terrain of Central India. Egyptian J. Remote Sens. Space Sci. **21**(1), 95–104 (2018)
23. Zidan, A., Abdalla, M., Khalaf, S., Saqr, A.M.: Kinetic energy and momentum coefficients for Egyptian irrigation canals. Mansoura Eng. J. **41**(1), 1–16 (2016). https://doi.org/10.21608/bfemu.2020.99368
24. Allam, A., Helal, E., Mansour, M.: Retarding contaminant migration through porous media using inclined barrier walls. J. Hydrol. Hydromechan. **67**(4), 339–348 (2019)
25. Mansour, M.M., Ellayn, A.F., Helal, E., Rashwan, I.M.H., Sobieh, M.F.: Delaying solute transport through the soil using unequal double sheet piles with a surface floor. Ain Shams Eng. J. **9**(4), 3399–3409 (2018). https://doi.org/10.1016/j.asej.2018.10.003
26. Saqr, A.M., Ibrahim, M.G., Fujii, M., Nasr, M.: Sustainable Development Goals (SDGs) associated with groundwater over-exploitation vulnerability: geographic information system-based multi-criteria decision analysis. Nat. Resour. Res. **30**(6), 4255–4276 (2021). https://doi.org/10.1007/s11053-021-09945-y
27. Saqr, A.M., Ibrahim, M.G., Fujii, M., Nasr, M.: Simulation-optimization modeling techniques for groundwater management and sustainability: a critical review. Adv. Eng. Forum **47**, 89–100 (2022). https://doi.org/10.4028/p-50l1j1
28. Saqr, A.M., Nasr, M., Fujii, M., Yoshimura, C., Ibrahim, M.G.: Monitoring of agricultural expansion using hybrid classification method in southwestern fringes of Wadi El-Natrun, Egypt: an appraisal for sustainable development. In: Ujikawa, K., Ishiwatari, M., Hullebusch, E.V. (eds) Environment and Sustainable Development. ACESD 2022. Environmental Science and Engineering. Springer, Singapore, (Forthcoming 2023)

Numerical Study on Response Characteristics of Layered Rocks Under Tunnel Blasting

Yingdong Pan[1], Shuyong Li[2](✉), Jian Guo[3], Feili Wu[2], and Xiaobin Ding[2]

[1] Key Laboratory of Transportation Tunnel Engineering, Ministry of Education, Southwest Jiaotong University, Chengdu 610031, China

[2] China Anneng Group Second Engineering Bureau Co., Ltd., Nanchang 330096, China
736766536@qq.com

[3] POWERCHINA Sichuan Electric Power Engineering Co., Ltd., Chengdu 610041, China

Abstract. Drilling and blasting method in tunnel may have some negative effects such as vibration and rock damage on environment. To understand the response characteristics of layered rock under tunnel blasting, a series of numerical tests are performed using 3DEC and the effects of thickness, occurrence and joint strength on the dynamic response of the layered rocks are analyzed and discussed. The results show the distribution shape of the plastic zone of layered rocks is mainly related to the occurrence of structural planes, while the distribution area is mainly affected by the thickness of the rock layers and the strength of the structural plane. Under the coupling of blasting stress and in-situ stress, the plastic zone of surrounding rock increases significantly and the plastic zone mainly develops along the direction that is perpendicular to the structural plane. With the increase of the layer thickness, the area of the plastic zone becomes larger under the coupling of blasting stress and in-situ stress. And with the increase of the inclination, the attenuation of peak vibration velocity at the measuring points above and below the tunnel decreases. The plastic zone of strong structural planes develops more, and with the enhancement of structural plane strength, peak vibration velocities at all measuring points increase but the attenuation of vibration velocity is basically unchanged.

Keywords: Layered rock · Tunnel blasting · Blasting response · Numerical simulation

1 Introduction

The layered sedimentary rocks are widely distributed in nature, accounting for about 2/3 of the total sedimentary rocks [1]. Layered rocks belong to transverse isotropic rocks, and the rock mass composition is basically the same in the direction parallel to the structural planes, while the direction perpendicular to the structural planes shows soft-hard alternation. And there are a large number of joints and fissures in the layered rock mass, resulting in obvious anisotropy of strength and deformability. When drilling and blasting method, the most widely used method in tunnel construction, is applied in the layered rocks, shock and vibration of blasting will easily lead to rock spalling and

X. Chen (Ed.): ICESE 2022, ESE, pp. 118–128, 2023.

overbreak due to the poor bonding force between the rock layers, which can seriously threaten the safety of personnel, delay the project progress, and cause economic loss [2]. Therefore, it is significant to study the response characteristics of layered rock under tunnel blasting.

In recent years, some scholars have conducted related studies. Chakraborty et al. [3] studied the effects of rock mass quality and joint orientation on tunnel blast performance by model tests. Ramulu et al. [4] presented a case study of tunnel construction project with regard to predicting and assessing blast-induced damage due to repeated dynamic loading. Deng et al. [5] studied damage of rock tunnel subject to blast-induced shock waves using UDEC software, and analyzed effects of joint properties on tunnel damage. Li et al. [6] investigated the influence of λ/D ratio on dynamic responses of underground tunnels under blasting loads using theoretical analysis and numerical simulation. Lisjak et al. [7] established a new modeling approach based on the combined finite–discrete element method to investigate the damage process and failure mechanisms around underground openings in clay shales. Mei [8] analyzed the mechanism of weak structural surfaces on contour formations and the causes of the severe overbreak, and proposed an optimization method of blasting parameters for large cross-section tunnel in horizontal layered rock. Li et al. [9] built physical models of an external crack around a tunnel (ECT) and conducted a series of blasting tests and analyzed the final failure modes of the ECT models based on stress wave theory. Zhang et al. [10] researched the influence of the dip angle of discontinuities and mechanical properties of weak seams on stratified rock mass stability using model tests and numerical simulation. Xie et al. [11] put forward a tension and compression-shear damage model and conducted a series of numerical simulations for damage evolution process and the damage zone induced by cut blasting.

However, the above studies paid less attention to dynamic responses of layered rocks with different properties under tunnel blasting. So, the aim of the present study is to explore response characteristics of layered rocks with different properties under tunnel blasting. In the paper, a series of numerical tests are performed using 3DEC and the response of layered rock under tunnel blasting are analyzed based on the effects of thicknesses, joint occurrences and strengths. The results including the distribution of plastic zone and the maximum vibration velocity are obtained, which provides a further understanding of tunnel blasting in layered rock.

2 Characteristics of Layered Rocks in the Tunnel Site

The Ganjialiang tunnel which is located in Xiahe County, Qinghai Province, is one of the tunnels in the Huangshengguan section of the new Xining—Chengdu line. The tunnel, with a total length of 10610 m, is an extra-long railway tunnel. The lithology of the zone which the tunnel passes is as follows: lower Triassic Jiangligou Formation and Shangaling Formation Slate, as shown in Fig. 1.

Based on the geological survey and the core samples in Fig. 2, the slates of the Lower Triassic Jiangligou Formation and Shangaling Formation have low rock strength and belong to grade IV and V surrounding rock. Its joints develop strongly and the thickness of slab is generally thin, and the cohesive force between the joints is poor and the slab

Fig. 1. Layered rocks in the tunnel site

Fig. 2. Typical rock cores

is easily separated into flakes or sheets. Under such conditions, the disturbance of blast load during excavation will easily lead to a series of problems such as crown collapse, face destabilization, large deformation and failure of initial support, which seriously threaten the tunnel construction. In order to reduce blast damage to the reserved rock, it is necessary to understand the response characteristics of different layered rocks under tunnel blasting.

3 Numerical Simulation

3.1 Calculation Model

The numerical model is based on the geological conditions of the Ganjialiang tunnel, and the depth of the tunnel is 400 m. It is assumed that the tunnel section is a straight-wall tunnel, and the height of the model is 80 m and the width is 80 m, and the height of the tunnel is 11 m and the width is 12 m. The mohr-Coulomb elastoplastic constitutive is used to the surrounding rock, and face-to-face contact is applied to the structural plane. The displacement constraints are set at the bottom and two sides of the model, and non-reflecting boundary conditions are set at four sides of the model, as presented in Fig. 3. In the figure, P represents the static load of ground stress and P_1 represents the dynamic load of tunnel blasting. The mechanical parameters of layered rocks are presented in Table 1.

Firstly, the model reaches the ground stress balance after 10 MPa stress is applied on the top of the model. Then, a tunnel is excavated in the model, and the blasting load, a triangular wave with a peak pressure of 560 MPa, is applied to the tunnel contour, as shown in Fig. 4. Finally, the dynamic response characteristics of tunnel under coupling of blasting stress and in-situ stress are analyzed.

Through 3DEC numerical simulation, the plastic zones of surrounding rock around the tunnel are obtained. Meanwhile, the peak vibration velocity in surrounding rock is monitored, and the measuring points are shown in Fig. 5.

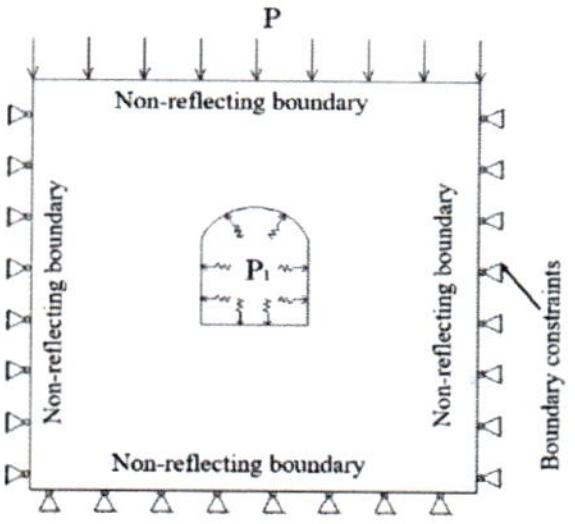

Fig. 3. The numerical model of tunnel blasting

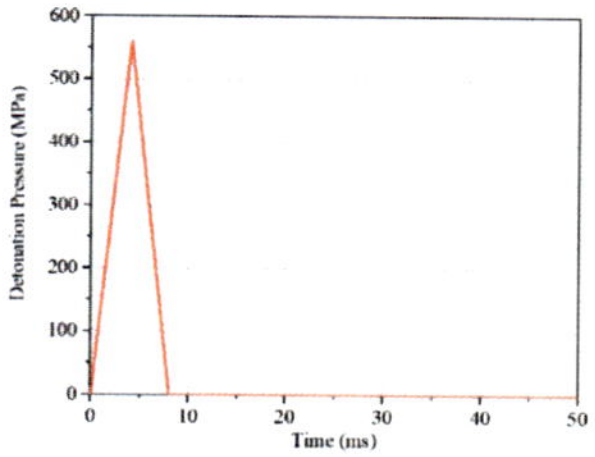

Fig. 4. The time-history curve of a triangular wave of blasting stress

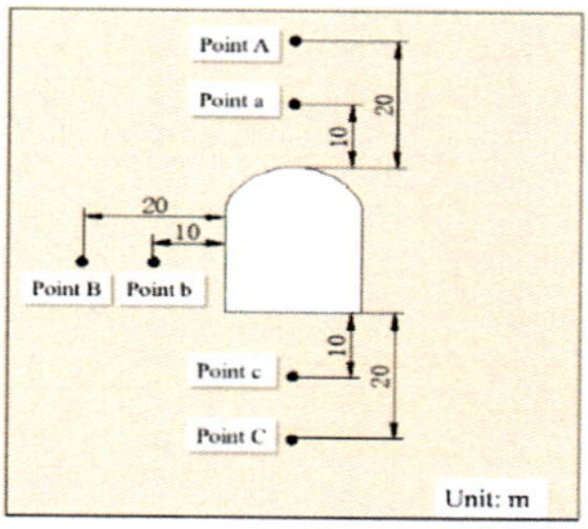

Fig. 5. The Location of measuring points for vibration velocity in surrounding rock

Table 1. Mechanical parameters of rock mass

Lithology	Density /kg·m^3	Bulk modulus/ GPa	Shear modulus/ GPa	Internal friction angle/°	Cohesive force/MPa	Tensile strength/MPa
Slab	2750	25	20	45	4	2

3.2 Experiment Cases

This paper focuses on the influence of the thickness of rock layers, occurrence and strength of structural plane on blasting response. Therefore, a series of experiments on the above three factors are carried out. Among them, the layered rock of medium strength and the occurrence of 0°and 0.5 m thickness is considered as a standard case to compare with other cases, as shown in Fig. 6(b).

Cases for Different Thicknesses of Rock Layers
According to the classification for the thickness of layered rocks in the "Railway Engineering Geology Manual", the layered rock masses are divided into four grades, including super-thick layer, thick layer, medium layer and thin layer. Based on the above standard, three kinds of typical thicknesses, including 0.3 m, 0.5 m and 1.0 m, are selected in the numerical simulation, as shown in Fig. 6.

Cases for Different Occurrences of Structural Planes
In order to study the influence of different occurrences of structural planes on tunnel blasting response, the cases for dip angles of 0°, 30°, 60° and 90° are selected, as shown in Fig. 7. It should be noted that the case for 0° is the standard case in Fig. 6(b).

Cases for Different Structural Plane Strengths
In order to study the influence of structural plane strength on the stability of surrounding rock, three typical structural plane strength parameters are taken for numerical simulation based on the "Engineering Rock Mass Classification Standard". The parameters of

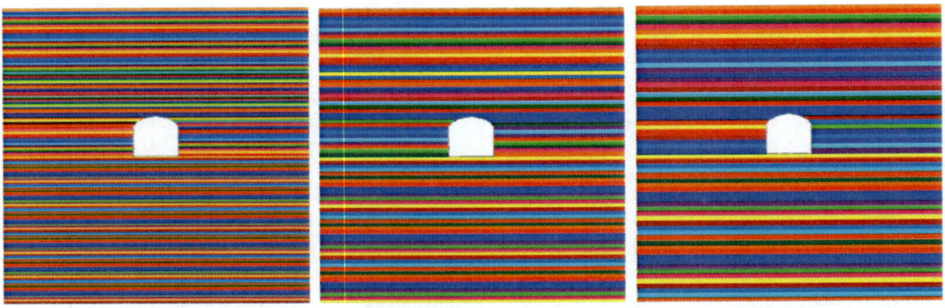

(a) 0.3 m thickness (b) 0.5 m thickness (c) 1.0 m thickness

Fig. 6. The models with different thicknesses of layered rocks

(a) 30° (b) 60° (c) 90°

Fig. 7. The models with different dip angles of structural planes

structural plane strength are shown in Table 2. It should be noted that the case of medium strength is the standard case in Fig. 6(b).

Table 2. Mechanical parameters of structural planes

Strength	Cohesive force /MPa	Internal friction angle /°	Tensile strength /MPa
Low	0.07	16	0.9
Middle	0.10	24	1.2
High	0.17	33	1.5

4 Results and Discussion

4.1 Effect of Thickness of Rock Layers on Response of Tunnel Blasting

Analysis of Plastic Zone Distribution

The plastic zone of surrounding rock after blasting load is applied is shown in Fig. 8. It can be seen that the plastic zone of surrounding rock develops in all directions, and the development range in the vertical direction is obviously larger than that in the horizontal direction. Comparing the area of plastic zone, the case for 1m is the largest, then is 0.5 m, and the least is 0.3 m. This is because the smaller thickness of layered rocks means the more amount of structural planes, and the more weak structural planes can absorb more explosion energy and reduce the propagation of explosion stress waves. Therefore, the area of plastic zone around the tunnel increased with the thickness of layered rock.

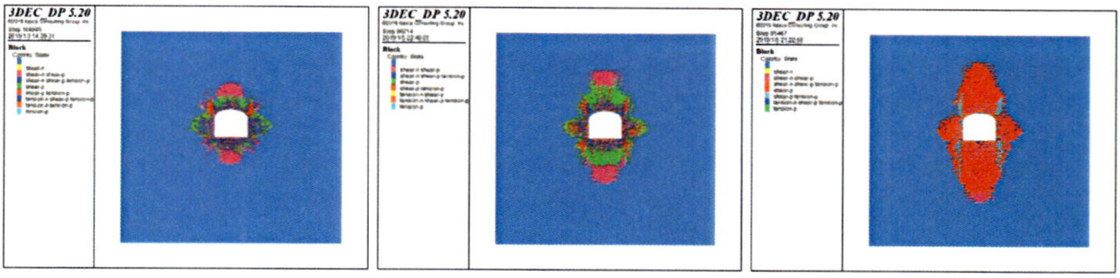

(a) 0.3 m thickness (b) 0.5 m thickness (c) 1.0 m thickness

Fig. 8. The plastic zones of surrounding rock with different thicknesses of layers after blasting

Analysis of Attenuation Law of Vibration Velocity

Table 3. Peak vibration velocities under different thicknesses of rock layers (unit: m/s)

Thickness	Monitoring point					
	a	A	b	B	c	C
0.3 m	0.597	0.209	0.696	0.404	1.002	0.398
0.5 m	0.822	0.313	0.921	0.510	1.494	0.523
1 m	1.037	0.38	1.077	0.541	1.739	0.582

As presented in Table 3, the peak vibration velocities of surrounding rocks with different strata thicknesses are extracted. At the same time, the attenuation of vibration velocity above the tunnel and on the left side of the tunnel and below the tunnel is shown as shown in Fig. 9. It can be seen that the maximum attenuation of vibration velocity is below the tunnel. In addition, with the increase of rock thickness, the attenuation of

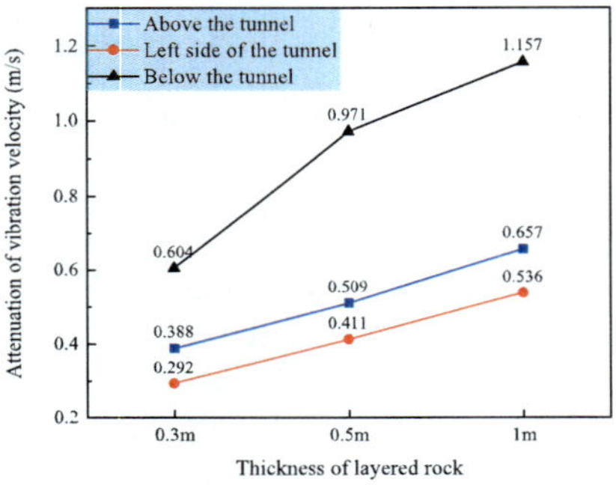

Fig. 9. Attenuation of vibration velocity under different thicknesses of rock layers

peak vibration velocity increases gradually, and the measuring points below the tunnel are the largest, followed by the measuring points above the tunnel and on the left side of the tunnel.

4.2 Effect of Occurrence of Structural Planes on Response of Tunnel Blasting

Analysis of Plastic Zone Distribution

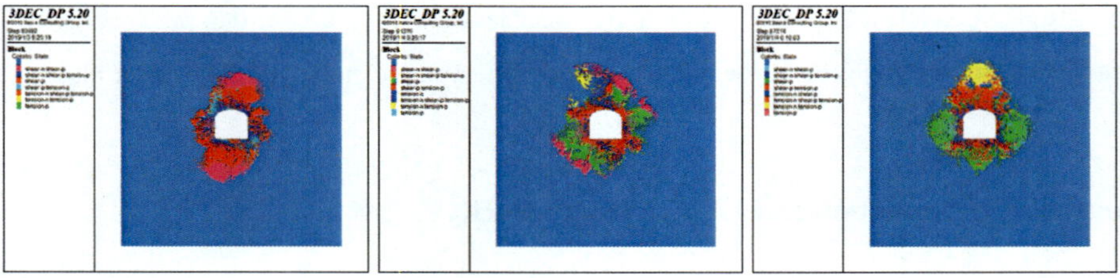

(a) 30° (b) 60° (c) 90°

Fig. 10. The plastic zones of surrounding rock with different occurrences of structural planes after blasting

When blasting stress is applied around the tunnel contour, the plastic zone of surrounding rock develops around the tunnel, and development is mainly perpendicular to the structural plane, as shown in Fig. 10. The above results show that the distribution of plastic zone is related to the occurrence of structural planes.

Analysis of Attenuation Law of Vibration Velocity

Through numerical tests, the dynamic responses of surrounding rock under different structural plane inclinations are obtained, and the maximum vibration velocities of each measured point are extracted in Table 4. Under the blasting stress, the particle vibration velocity time history above the tunnel crown, the left sidewall of the tunnel and the floor of the tunnel with different dip angles, as shown in Fig. 11. It can be seen that the attenuation amplitude of peak vibration velocity on the measuring points above the tunnel crown and below the tunnel floor decreases with the increase of the dip Angle. On the left side of the sidewall, the attenuation amplitude of particle peak vibration velocity reaches the maximum at the inclination angle of 90°.

Table 4. Peak vibration velocities under different occurrences of structural planes (unit: m/s)

Angle	Monitoring point					
	a	A	b	B	c	C
0°	0.822	0.313	0.921	0.510	1.494	0.523
30°	0.653	0.239	0.714	0.261	1.162	0.387
60°	0.5	0.241	0.903	0.334	0.773	0.306
90°	0.771	0.581	1.08	0.438	1.044	0.622

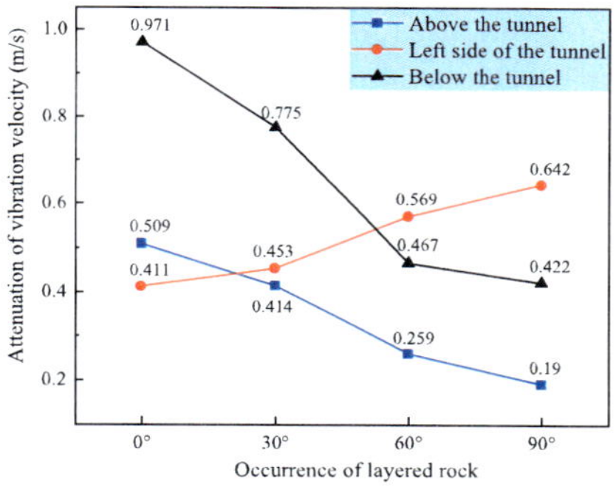

Fig. 11. Attenuation of vibration velocity under different occurrences of structural plane

4.3 Effect of Strength of Structural Planes on Response of Tunnel Blasting

Analysis of Plastic Zone Distribution

When blasting stress is applied around the tunnel contour surface, the development of plastic zone is mainly perpendicular to the structural plane, as shown in Fig. 12. Comparing the development range of the plastic zone, the plastic zone of high strength structural plane is a little larger than that of medium strength structural plane and that of low strength structural plane. This is because the higher strength of the structural plane means the better bond performance, and the condition is more conducive to the propagation of blasting stress waves in the surrounding rock. So, the area of the plastic zone around the tunnel is larger.

Analysis of Attenuation Law of Vibration Velocity

Through numerical tests, the peak vibration velocity of surrounding rock under different

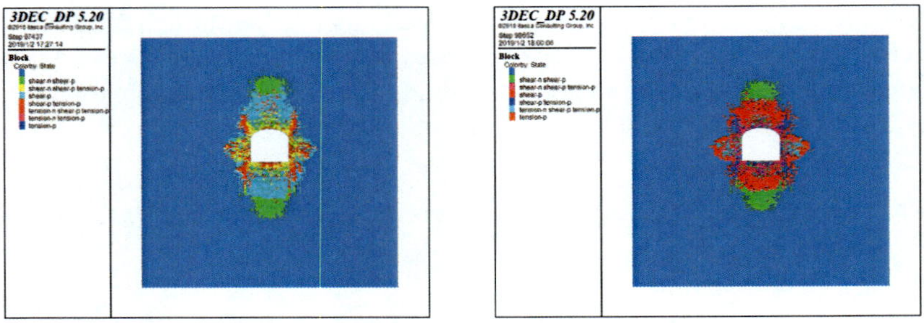

(a) High-strength joints (b) Low-strength joints

Fig. 12. The plastic zones of surrounding rock with different structural plane strengths after blasting

structural plane strengths is obtained, as presented in Table 5. Furthermore, the attenuation law of vibration velocities above the tunnel crown, on the left of the tunnel and below the tunnel with structural plane strengths are shown in Fig. 13. It can be seen that during tunnel blasting, the maximum vibration velocity is below the floor, and the corresponding attenuation value is also the largest. With the enhancement of structural plane strength, peak vibration velocities at all measuring points increase but the attenuation of vibration velocity is basically unchanged.

Table 5. Peak vibration velocities under different structural plane strengths (unit: m/s)

Strength	Monitoring point					
	a	A	b	B	c	C
Low	0.799	0.303	0.89	0.479	1.433	0.469
Middle	0.822	0.313	0.921	0.510	1.494	0.523
High	0.957	0.4	0.956	0.549	1.541	0.584

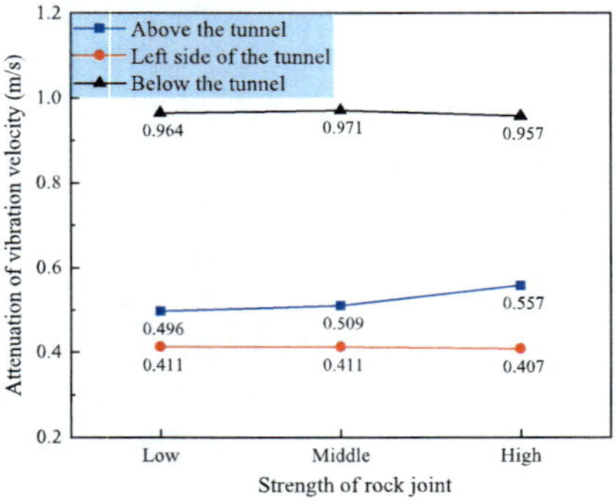

Fig. 13. Attenuation of vibration velocity under different structural plane strengths

5 Conclusion

In this paper, we conducted a series of numerical tests using 3DEC to investigate the dynamic responses of layered rocks under tunnel blasting. The conclusion can be addressed as follows:

(1) The distribution shape of the plastic zone of layered rocks is mainly related to the occurrence of structural planes, while the distribution area is mainly affected by the thickness of the rock layers and the strength of the structural planes.
(2) Under the coupling of blasting stress and in-situ stress, the plastic zone of surrounding rock increases significantly and the plastic zone mainly develops along the direction that is perpendicular to the structural plane.
(3) With the increase of the rock thickness, the area of the plastic zone became larger under the coupling of blasting stress and in-situ stress. The reason is the smaller thickness of layered rocks means the more amount of structural planes, and the more weak structural planes can absorb more explosion energy and reduce the propagation of explosion stress waves.
(4) The occurrence of structural plane plays a decisive role in the distribution of the plastic zone of surrounding rock. The attenuation of peak vibration velocity at the measuring points above and below the tunnel decreases with the increase of the inclination. In the case of 0°, a bending fracture occurred at the crown after blasting load is applied. In the case of 60°, the right spandrel of the tunnel is prone to fracture and over-excavation. Therefore, reducing the explosive amount or increasing the coefficient of decoupling charge can improve the excavation contour of tunnel.
(5) The plastic zone of high strength structural plane is a little larger than that of medium strength structural plane and that of low strength structural plane. And with the enhancement of structural plane strength, peak vibration velocities at all measuring points increase but the attenuation of vibration velocity is basically unchanged.

References

1. Wu, B.: Study on disturbance zone evolution and anchoring mechanism of surrounding rock in layered rock tunnel. China University of Geosciences, Wuhan (2016)
2. Yang, J.H., Lu, W.B., Hu, Y.G., et al.: Numerical simulation of rock mass damage evolution during deep-buried tunnel excavation by drill and blast. Rock Mech. Rock Eng. **48**(5), 2045–2059 (2015)
3. Chakraborty, A.K., Jethwa, J.L., Paithankar, A.G.: Effects of joint orientation and rock mass quality on tunnel blasting. Eng. Geol. **37**(3–4), 247–262 (1994)
4. Ramulu, M., Chakraborty, A.K., Sitharam, T.G.: Damage assessment of basaltic rock mass due to repeated blasting in a railway tunnelling project–A case study. Tunn. Undergr. Space Technol. **24**(2), 208–221 (2009)
5. Deng, X.F., Zhu, J.B., Chen, S.G., et al.: Numerical study on tunnel damage subject to blast-induced shock wave in jointed rock masses. Tunn. Undergr. Space Technol. **43**, 88–100 (2014)

6. Li, C.J., Li, X.B.: Influence of wavelength-to-tunnel-diameter ratio on dynamic response of underground tunnels subjected to blasting loads. Int. J. Rock Mech. Min. Sci. **112**, 323–338 (2018)
7. Lisjak, A., Grasselli, G., Vietor, T.: Continuum–discontinuum analysis of failure mechanisms around unsupported circular excavations in anisotropic clay shales. Int. J. Rock Mech. Min. Sci. **65**, 96–115 (2014)
8. Mei, J., Zhang, W.Z., Xu, B.S., et al.: Optimization methods of blasting parameters of large cross-section tunnel in horizontal layered rock mass. Geotech. Geol. Eng. **39**(7), 5309–5323 (2021)
9. Li, X.H., Zhu, Z.M., Wang, M., et al.: Influence of blasting load directions on tunnel stability in fractured rock mass. J. Rock Mech. Geotech. Eng. **14**(2), 346–365 (2022)
10. Zhang, Z.X., Xu, Y., Kulatilake, P.H.S.W., et al.: Physical model test and numerical analysis on the behavior of stratified rock masses during underground excavation. Int. J. Rock Mech. Min. Sci. **49**, 134–147 (2012)
11. Xie, L.X., Lu, W.B., Zhang, Q.B., et al.: Damage evolution mechanisms of rock in deep tunnels induced by cut blasting. Tunn. Undergr. Space Technol. **58**, 257–270 (2016)

Clean Energy Technology and Building Energy Management

Prediction and Regression Analysis of Trans-Himalaya New Energy Development Trends Based on Logistic Model

Panpan Deng(✉)

Leshan Vocational and Technical College, Leshan 614000, China
20764775@qq.com

Abstract. The trans-Himalaya region has a large population, a large demand for energy, and a high dependence on foreign energy supply, and its energy security and economic development face severe challenges. This paper focuses on exploring a new path of high-quality coordinated development serving Sichuan photovoltaic industry advantages across the Himalayan region, forming the industrial layout concept of "two poles together" and the innovative development mechanism of "one body and one institute". Finally, the Logistic model was used to predict the trend of new energy development and energy consumption across the Himalayas, and to analyze the comprehensive optimization index. The results of the regression analysis confirm the feasibility of the industrial layout and the innovative development mechanism, which is conducive to all countries in the region to actively promote the energy transformation and promote the economic and social development of all countries.

Keywords: Trans-Himalaya region · Two Poles One Belt · One Mechanism One College · Logistic model · Comprehensive optimization

1 Background

In order to better build a new development pattern with domestic cycle as the main body and domestic and international cycles promoting each other, in the "new track" of promoting the high-quality development of local industries, various regions have gradually promoted the research of trans-Himalaya exchanges and cooperation. Literature [1] studied that Tibet has strengthened cooperation with Nepal on clean energy, especially hydropower development. Literature [2] studied the development and investment opportunities of key industries in Thailand. Literature [3] explores the historical traceability, realistic orientation and path of the cultural exchanges between Qinghai and Nepal. Literature [4] taked a sample of five major economies across the Himalayan region, estimated its financial, economic, political and integrated investment environment index based on the ICRG data of PRS Group ICRG data, and empirically analyzed the impact of the investment environment in the region on INFDI. Literature [5] mentioned that a vocational college in the western region made use of its special geographical advantages

X. Chen (Ed.): ICESE 2022, ESE, pp. 131–143, 2023.

and made full use of the existing teaching resources to carry out the health cooperation between Bangladesh, China, India and Myanmar. Literature [6] taked a sample of five important economies across the Himalayan region to analyzed the interaction between the investment environment, direct investment flow and economic development in the region. There are many studies on the cooperation between China and trans-Himalaya countries in the literature, involving economy, trade, tourism, culture, energy and other fields. However, the research on the energy cooperation between China and countries in the region is relatively insufficient. Overall, the current domestic and foreign about across the Himalayan research has made great progress, but most research focus on political, economic, ecological, security and other fields, to research in the field of energy is relatively insufficient, systemic and sustainability also lack, especially across the Himalayas as a whole regional energy cooperation research is less.

Under the background of economic globalization and the gradual deepening of the "Belt and Road" international cooperation initiative [7], this paper focuses on exploring the construction and construction of the cooperative industry layout and innovative development mechanism with the new energy base in Sichuan as a pilot. The Logistic model will be applied to predict the trend of new energy development and energy consumption across the Himalayan region. And it aslo will be analyzed by using the comprehensive optimization index. In this way, we will give full play to our advantages in clean energy resources and the foundation of industrial development, and foster new green and low-carbon drivers that support high-quality development [8].

2 Cooperation Foundation

Consensus on energy cooperation among most countries across the Himalayas has been strengthened. Under the background of the conflict between Russia and Ukraine, the fossil energy crisis is constantly prominent, the potential advantages of renewable energy development are constantly emerging, and the regional energy development and cooperation have a significant cooperation consensus. In particular, photovoltaic power generation is becoming the most competitive form of power supply in more and more countries. It is expected that the global photovoltaic market will continue to maintain high growth in the future. Driven by favorable prospects and platform support, global economic recovery and other favorable factors, the global new photovoltaic installed capacity is growing rapidly. It is expected that the global annual new photovoltaic installed capacity will reach 250GW from 2022 to 2025.

2.1 Consensus on Energy Cooperation was Initially Reached

In the post-epidemic era, trans-Himalaya countries jointly face multiple problems, such as energy shortage, economic recovery and dual-carbon targets. In this context, the only way for countries across the Himalayas is to build a community with a shared energy destiny with new energy development and jointly solve several key issues. In addition, the global climate warming caused by carbon emissions makes it very necessary for countries around the world to realize the overall progress of environmental governance.

2.2 The Advantages of Photovoltaic Cooperation are Increasingly Emerging

In serving the overall national "double-carbon" strategy, Sichuan has actively participated in the construction of a new development pattern with the domestic big cycle as the main body and the domestic and international double cycles promoting each other. Sichuan province takes crystalline silicon photovoltaic as a "key project" of a strong industrial province, and develops the crystalline silicon photovoltaic industry with a long history, rich resources and solid foundation. It focuses on improving the industrial chain, expanding the scale of advanced production capacity, and consolidating the leading position in the industry, and has built a world-class crystalline silicon photovoltaic industrial base "China Green Silicon Valley" industrial park.

The trans-Himalaya region is complementary to Sichuan in terms of photovoltaic energy distribution, economic structure and geographical types, thus having potential complementary advantages in energy cooperation. Due to the dominant position in the development of traditional energy and many bottlenecks in the development of renewable energy, this advantage has not been highlighted in the previous independent development of all countries across the Himalayan region. Under the influence of COVID-19 and energy security, the instability of traditional energy has begun to become prominent. Under the condition of technical conditions, gradually adopting more renewable energy is the best choice for countries in the region.

3 Build "Two Poles One Belt"

The concept of an energy community with a shared future is based on the concept of a community with a shared future for mankind, and builds an industrial layout concept with green and sustainable development as the core. This is in line with the trend of clean, green and low-carbon development in today's world, and meets the needs to promote the transformation of regional energy structure in the face of the epidemic [9].

"Two poles" fouces on Sichuan-India as the core of the trans-Himalaya green photovoltaic economic cooperation belt core areas [10]. Sichuan province regards crystalline silicon photovoltaic as a "key project" of a strong industrial province, and has a long history, rich resources and a solid foundation for developing the crystalline silicon photovoltaic industry. Sichuan province has made great efforts in improving the industrial chain, expanding the scale of advanced production capacity, and consolidating its leading position in the industry, and has built a world-class "China green silicon Valley" industrial park of crystalline silicon photovoltaic industrial base. At present, half of the world's top 10 photovoltaic enterprises have settled in the park. By 2025, "China Green Silicon Valley" will form a production scale of 800,000 tons of silicon, 100GW pull rod cutting side, 100GW slices, 50GW cells and 30GW modules. The output value of crystalline silicon photovoltaic will exceed 150 billion yuan, forming a world-class crystalline silicon photovoltaic industry base. As the world's fifth largest economy, India has diversified economy and industries, perfect infrastructure, detailed geographical advantages and rich solar energy resources. India's information technology industry is leading in the world. With the help of the early advantages of Sichuan crystal silicon photovoltaic industry, India can jointly explore the establishment of dual-carbon target trading mode and trading strategies in the countries in the region. In recent years, many photovoltaic enterprises in the "China Green Silicon Valley" industrial park in Sichuan province have established cooperation with India and the intention to settle down in India. Jinkosolar ranks first in the photovoltaic module market in India, and Tongwei company is a major supplier of cells in India. GCL group and SoftBank Investment jointly invested us $930 million in a joint venture in India for the main photovoltaic business. Under the influence of policy guidance and market demand, the future development trend of centralized and distributed photovoltaic power generation is more obvious. According to the total new energy production statistics in Table 1, China ranks first. Among the top countries, the trans-Himalaya region occupy four.

Table 1. Total renewable energy production

Ranking	Country	quadrillion Btu
1	China	21.615
2	United States	20.084
3	Brazil	5.486
4	France	5.078
5	Canada	4.906
6	Russia	3.861
7	India	3.362
8	Germany	3.03
9	Japan	2.284
10	South Korea	1.615
11	United Kingdom	1.59
12	Sweden	1.553
13	Spain	1.525
14	Turkey	1.171
15	Norway	1.157
16	Ukraine	0.963
17	Italy	0.909
18	Indonesia	0.707
19	Venezuela	0.642
20	Belgium	0.629
21	Switzerland	0.623
22	Mexico	0.597
23	Vietnam	0.541

Data Source: www.eia.gov.

"One Belt" is mainly held in Pakistan, Thailand, Vietnam and other countries along the Himalayas. Under the guidance of the "two poles", we will integrate the existing industrial development resources of various countries, find the right positioning, conform to the general trend of green and low-carbon development, and focus on developing green industries. Take cells as an example. From the perspective of the current production layout of the photovoltaic industry, combined with the advantages of upstream and downstream supply chain supporting, production factor cost, labor, local industrial policies and other aspects, the production layout of cells in 2021 is mainly concentrated in Asia, with a production capacity of about 422.2GW, accounting for 99.7% of the global total production capacity. By the end of 2021, the battery production capacity layout of leading photovoltaic enterprises in Malaysia, Vietnam, Thailand and other countries has been initially formed, reaching 31.7GW. As shown in table 2, Chinese photovoltaic enterprises have a large advantage in the early layout, which has gradually highlighted the strong momentum of joint development [11].

Table 2. Production capacity layout of photovoltaic enterprises in the trans-Himalaya region

country	enterprise	Annual Capacity in 2021 (MW)
Malaysia	JinKo	6300
	Longi	3800
	Hanwha Group	3000
	Huasun	1800
	JA Solar	1400
Vietnam	Longi	5000
	Trina S olar	4800
	Boviet Solar	1500
Thailand	Canadian Solar	4200
	Ventec	1660
	Trina S olar	1200
	CHNT	1000

Data Source: CPIA, 2022.3

4 Build "One Mechanism One College"

The innovative development mechanism of "One Mechanism One College" focuses on solving the technical bottleneck and the shortage of technical skills and personnel in the trans-Himalaya green photovoltaic economic cooperation belt. "One Mechanism" is a trans-Himalaya green and low-carbon cooperation community [12]. With "One Mechanism" as the overall deployment, through the government building a platform, the new energy industry leads, various enterprises deeply participate, universities and research institutes jointly build a photovoltaic technology service platform, and implement the green enterprise leading and clean energy contribution action. Focusing on high standards, we will work together in technological innovation, project construction, and environmental protection. Focusing on improving people's lives, adhering to the concept of a community with a shared future for mankind, adhering to the responsibilities of a major country in our business, and improving people's well-being. Focus on sustainability, inject more advanced technologies into the "Two Poles One Belt", and reflect the new concept of mutually beneficial and coordinated development of a green and low-carbon cooperative community. In terms of financial cooperation platforms, regional financial cooperation platforms such as the Asian Infrastructure Investment Bank can provide financial support for regional energy cooperation. Various existing regional cooperation platforms provide broad exchange platforms, financial support and institutional experience for promoting regional energy development and cooperation, and play an important role in promoting regional energy infrastructure construction, building an interconnected energy transportation network, promoting technology and talent exchanges, and ensuring energy security. Based on the energy cooperation projects in the region, we will innovate and practice the energy + cooperation mechanism. We will give full play to the experience and advantages of Sichuan in infrastructure construction and epidemic prevention and control, and, with the help of the implementation and promotion of energy projects, carry out more mutual assistance and cooperation in road construction, power network construction, photovoltaic poverty alleviation, and epidemic prevention and control management.

Relying on multinational enterprises and universities with international experience in exchange and cooperation, "One College" will jointly build a trans-Himalaya Collaborative Photovoltaic Industry College. The key to economic poverty and energy poverty across the Himalayas, and the key to economic development and energy transformation is the lack of talent support. Through the innovation of the school organization mechanism and system, the participants can participate in the all-round, deep and more active aspects of personnel training. Optimize the discipline structure around the industrial structure, promote the division of labor and cooperation of "education chain-industrial chain-talent chain-innovation chain", become a new carrier for serving the transformation and upgrading of the photovoltaic industry across the Himalayan region, improve the hematopoietic capacity of "Two Poles One Belt", and continuously provide the required technical and skilled talents for the industrial development.

5 Logistic Regression Analysis

New energy power generation is becoming the most competitive form of power supply in more and more countries, and it is expected that the global new energy market will still maintain rapid growth in the future. This section mainly predicts and analyzes the total new energy production of trans-Himalaya countries in the next 50 years. The conditions that affect the total amount forecast are always changing, which can be summed up as the following points.

(1) Energy composition, energy consumption structure and supply and demand situation of each country.

(2) Economic factors, including GDP, industrial structure, per capita disposable income, etc., should be considered in the annual forecast due to the continuity of a region's continuous economic development.
(3) Social infrastructure construction, including power grid construction, transportation facilities construction, etc.
(4) Population situation.
(5) Meteorological conditions, especially temperature and irradiance, have an obvious impact on photovoltaic.
(6) Competitiveness of other energy sources. Different energy technologies are competitive and alternative.
(7) Random factors. Other large number of random factors that can cause energy consumption, including energy policy, weather mutations, major events, etc.

Refer to the early development of energy consumption and production trend of developed countries. Generally, the growth rate is slow in the early stage, the medium-term development is fast, and the gentle growth in the later stage, and finally reaches a state of saturation. The developmental trend has similarities to the Logistic curve model. Therefore, the Logistic function is used for the prediction analysis in this paper.

Mathematical model of the Logistic curve

$$y = \frac{c}{1 + e^{a-b\cdot t}} \tag{1}$$

In formula, y is the total amount of energy. c is the function saturation value. a is the function initial value parameter, and b is the growth parameter. T stands for time.

The accuracy of the prediction results in the prediction analysis is explained by the error, and the relative error is mainly used in this paper. In the following formula, the y_i represents the actual value, and $\overline{y}_i$ represents the predicted value.

$$\partial = \frac{\left|y_i - \overline{y}_i\right|}{y_i} \times 100\% \tag{2}$$

To analyze the competitiveness evaluation of certain energy applications, the comprehensive optimization index can be selected for analysis.

$$M_{i-1} = \frac{k_{i-1} \times M_i}{k_i} \tag{3}$$

where, M_{i-1} is the comprehensive optimization index of new energy output in last year. M_i is the comprehensive optimization index of new energy output in this year. k_{i-1} is the proportion of new energy in the total energy consumption in last year. k_i is the proportion of new energy in the total energy consumption in this year.

Representative China and India across the Himalayan region were selected as the pilot analysis objects. The data of total energy consumption and total new energy production from 1980 to 2019 was collected, and the data from 1980 to 2010 was taken as the basic data, and the Logistic curve was selected to analyze and predict the data from 2011 to 2019, and the accuracy of the method was analyzed with the relative error as

Table 3. Energy consumption forecast for China

Years	Predicted Value (quad Btu)	A ctual Value (quad Btu)	Relative Error (%)
2011	125.32	122.34	2.44%
2012	133.24	130.93	1.76%
2013	138.33	135.51	2.08%
2014	140.67	137.54	2.27%
2015	144.21	137.01	5.25%
2016	146.78	137.85	6.48%
2017	148.98	142.25	4.73%
2018	154.11	147.25	4.66%
2019	157.44	151.84	3.69%

Data Source: www.eia.gov.

Table 4. New energy output forecast for China

Years	Predicted Value (quad Btu)	A ctual Value (quad Btu)	Relative Error (%)
2011	7.98	7.78	2.57%
2012	9.99	9.59	4.17%
2013	10.46	10.66	1.88%
2014	12.38	12.35	0.24%
2015	14.12	13.12	7.62%
2016	14.72	14.22	3.52%
2017	15.63	15.53	0.64%
2018	16.98	16.78	1.19%
2019	19.15	18.05	6.10%

Data Source: www.eia.gov.

the judgment standard. The forecast results of total energy consumption and total new energy production are shown in the Tables 3, 4, 5 and 6.

According to the forecast results, using the same Logistic curve, the total energy consumption and total new energy production of China and India are somewhat different from the actual value. Relative error was used to characterize the accuracy of the predictions by each year. The results show that the relative error was all controlled within 10%. This result has high reference value for long-term prediction and belongs to high-precision prediction.

Further medium to long term prediction analysis in more countries across the Himalayas using Logistic curve models. Seven countries were selected from China,

Table 5. Energy consumption forecast for India

Years	Predicted Value (quad Btu)	A ctual Value (quad Btu)	Relative Error (%)
2011	22.56	23.11	2.40%
2012	23.78	24.28	2.04%
2013	24.67	24.95	1.11%
2014	27.87	26.62	4.68%
2015	28.88	27.31	5.76%
2016	29.95	28.25	6.03%
2017	30.99	29.47	5.16%
2018	33.47	31.28	6.99%
2019	34.87	31.90	9.31%

Data Source: www.eia.gov.

Table 6. New energy output forecast for India

Years	Predicted Value (quad Btu)	A ctual Value (quad Btu)	Relative Error (%)
2011	1.66	1.70	2.07%
2012	1.71	1.59	7.56%
2013	1.8	1.87	3.56%
2014	1.96	1.90	3.42%
2015	1.98	1.94	2.14%
2016	2.06	2.10	1.98%
2017	2.31	2.35	1.87%
2018	2.64	2.57	2.79%
2019	2.97	2.85	4.25%

Data Source: www.eia.gov.

India, Malaysia, Nepal, Pakistan, Thailand and Vietnam. The data of total energy consumption and total new energy production of various countries from 1980–2019 is extracted, and to forecast the data trend of 2020–2050.

According to the forecast results in Fig. 1–2, the global economic growth has slowed due to the COVID-19 pandemic. As a result, the energy development in the trans-Himalaya region is also threatened to a certain extent, and the economic recovery and energy development are facing uncertainties. The energy consumption in the trans-Himalaya region is dominated by traditional energy consumption, but due to the lack of traditional energy consumption in the region, the traditional energy consumption relies on imports for a long time, and it has a high dependence on foreign countries.

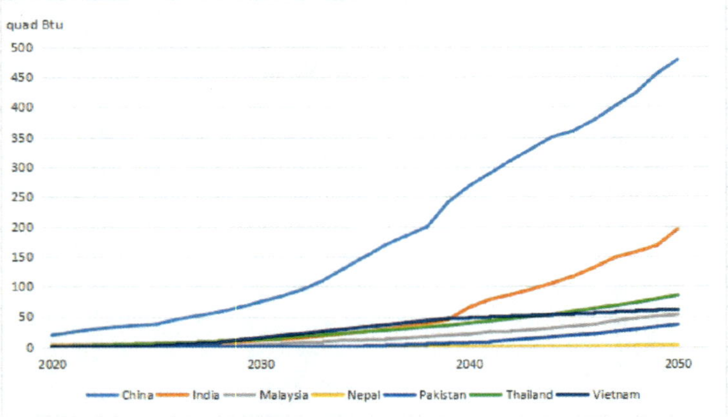

Fig. 1. New energy output forecast

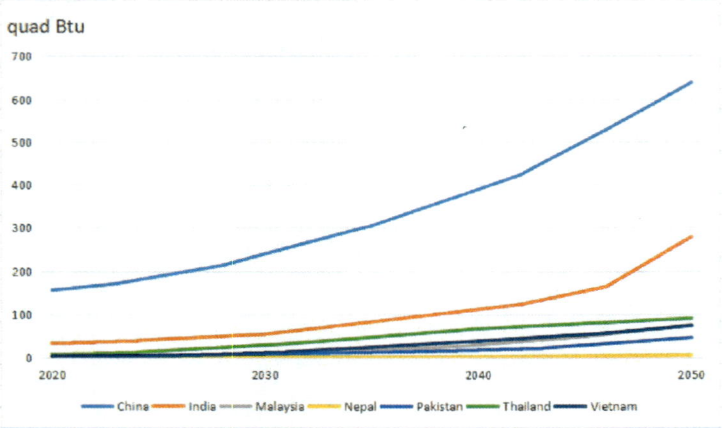

Fig. 2. Energy consumption forecast

China's new energy development started relatively early. Sichuan Province will take the pilot to build a new ecology of "Two Poles One Belt", give full play to the integrated incentive benefits, effectively stimulate the development vitality of Sichuan leading the green and low-carbon industries, and promote the core competitiveness of China's new energy. At the same time, we will vigorously develop green leading industries and form a agglomeration development trend and innovative development mechanism in the trans-Himalaya region. In 2020–2025 stage, with new energy development as an opportunity to domain countries focus on renewable energy and new energy cooperation, relying on the countries in the domain of renewable energy resources, development technology, financial and market complementarity, together to promote the outbreak of regional energy structure transformation and optimization, to form a new mode of green, sustainable energy cooperation. Gradually stimulate and drive the economic recovery of all countries in the region. According to the forecast results, national energy consumption will gradually start to grow around 2030. This has not only revitalized the regional energy industry, but also promotes the needs of energy transformation and economic development. As can be seen from Fig. 3, by around 2050, except for Nepal, the new energy output of other countries accounted for more than 60% of its total energy consumption. Among them, Thailand accounted for 92% of its natural environmental resources and industrial development advantages such as sufficient sunshine. Thus, the building of a trans-Himalaya green and low-carbon cooperative community and promoting the construction of a new fair, efficient, open and green energy governance system will greatly promote the achievement of the dual-carbon goal.

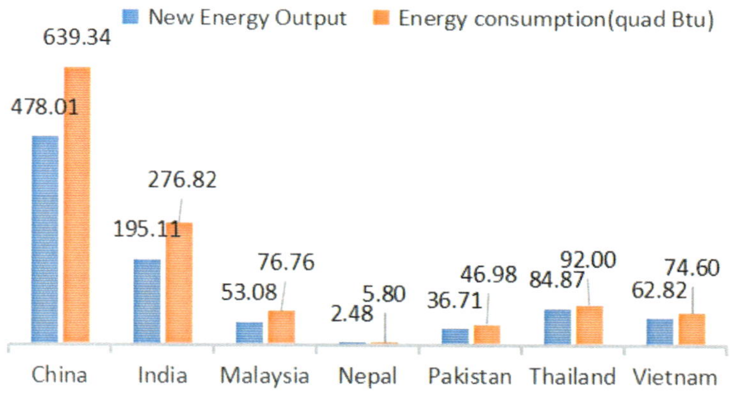

Fig. 3. Share of new energy output in energy consumption in 2050

Because the comprehensive energy consumption volume of China and India is large, it is impossible to intuitively compare and analyze the advantages of new energy development with other countries. The next step is to analyze the comprehensive optimization of new energy output in various countries. The initial comprehensive optimization value of each country in 1980 is 1, which is available as calculated by formula (3). From 1980 to 2050, the comprehensive optimization value of new energy output in each country is shown in Fig. 4.

According to the comprehensive optimization analysis, it can be seen that before 2020, due to the overall slow development of the new energy industry, the gap between countries is not big, but later on, China, Thailand, Malaysia and Vietnam will have a strong development momentum. Thailand, in particular, has the first place in overall excellence. This is consistent with the conclusions of the previous analysis. India is a large size, but its own industrial layout is not perfect. The development of India's new energy industry needs China to rely on a more perfect industrial chain system as a strong support to play a better role, which also confirms the rationality of the previous "Two Poles" industrial layout.

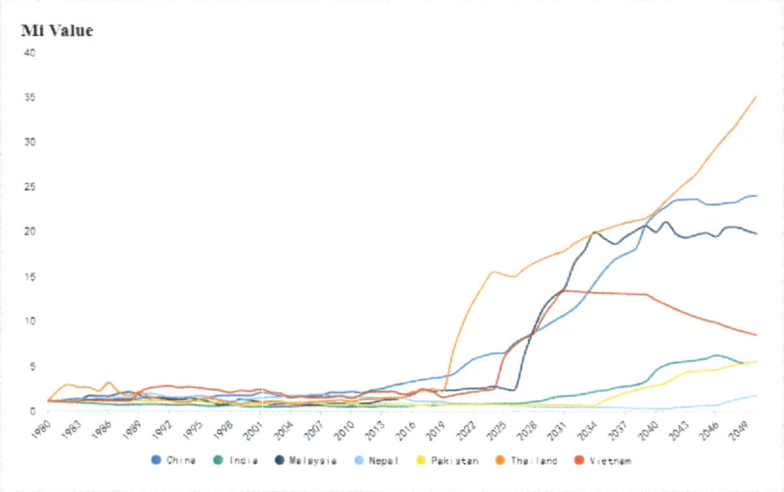

Fig. 4. M_i index analysis

6 Conclusion

Under the COVID-19 epidemic, the international energy pattern and world economic development have been severely impacted, and the global energy governance disorder has intensified. The trans-Himalaya region has a large population, large energy demand and high dependence on energy supply, and energy security and economic development face severe challenges. At present, through the reasonable layout of the pilot, the advantages of Sichuan photovoltaic industry have been transformed into a new path of high-quality coordinated development serving the Himalayan region, and the industrial layout concept of "two poles together" and the innovative development mechanism of "one body and one institute" have been formed. Taking the opportunity of new energy cooperation will help countries in the region strengthen mutual trust, deepen cooperation consensus and get rid of the economic downturn in the post-epidemic era. By establishing a Logistic function model to analyze the trend of future energy consumption and new energy output in countries across the Himalayas, it shows the feasibility of the industrial layout and innovative development mechanism. All Countries in the region should actively promote the energy transformation, jointly cope with the energy crisis and economic development difficulties, jointly protect the regional energy security, and jointly promote the economic and social development of all countries.

Acknowledgments. Foundation Project: Supported by Center for trans-Himalaya Studies project (No. KX2022C03).

References

1. Xiaoyang, X.: Research on trans-Himalaya hydropower development cooperation in China (Tibet) and Nepal under the dual-carbon Goal. Tibet Dev. Forum **06**, 30–36 (2021)
2. Penghui, F., et al.: Development and investment opportunities of key industries in Thailand. Int. Econ. Cooper. **10**, 56–60 (2016)
3. Faping, S., Fang, H.: Research on the path and countermeasures of deepening the cultural exchanges between Qinghai and Nepal under the background of "Belt and Road." Gansu CPPCC **05**, 56–60 (2021)
4. Junrong, L.: trans-Himalaya regional investment environment index estimates and the impact on INFDI the is based on an empirical analysis of the five major economies in the region. J. Guizhou Univ. Finan. Econ. **06**, 26–35 (2018)
5. Yingxin, D.: Strengthen cross-border cooperation in the field of health and disease control to improve the health level of people in the trans-Himalaya border areas. Sci. Chin. **09**, 124–125 (2016)
6. Ying, J., Lei, Z., Junrong, L.: Analysis of the Interaction between trans-Himalaya economic development and direct investment. Stat. Dec. Making **35**(15), 146–149 (2019). https://doi.org/10.13546/j.cnki.tjyjc.2019.15.034
7. Arase, D., Miguel, P., de Medeiros Carvalho, A.R.: The Belt and Road Initiative in Asia, Africa, and Europe. Taylor and Francis 26 May 2022
8. Anonymous. TEX-ISLE: Assessment Affirms Leadership Position in Low-Carbon and High ESG Performance. Oil & Gas J. **120** (3) (2022)
9. Li ,Y., Lan, S., Ramirez-Javier, P., Wang, X.: Achieving a low-carbon future through the energy-chemical nexus in China (vol 4, pg 6141, 2020). Sustainable Energy & Fuels, 5 (18) (2021)
10. Iqbal, S.I., Zargar, S.A., Ganie, A.H., Shah, M.A.: Distribution dynamics of Arnebia euchroma (Royle) I.M.Johnst. and associated plant communities in trans-Himalaya Ladakh region in relation to local livelihoods under climate change. Trees, Forests and People, 7 (2022)
11. Marifatul, H.S., et al.: Traditional Usage of Wild Fauna among the Local Inhabitants of Ladakh, trans-Himalaya Region. Animals **10**(12) (2020)
12. Science Earth Science. Findings on Earth Science Discussed by Investigators at Bose Institute (Geomicrobial Dynamics of trans-Himalaya Sulfur-borax Spring System Reveals Mesophilic Bacteria's Resilience To High Heat). Ecology Environment & Conservation, 2020

Algae Biomass for Biofuel Production; Yield Analysis of *Chlorella Vulgaris* and *Scenedesmus Meyen* in Different Culture Media

Mona Alariqi[1], Wei Long[1], and Baba Imoro Musah[2(✉)]

[1] School of Economics, Department of Industrial Economics, Wuhan University of Technology, Wuhan, China
[2] School of Resources and Environmental Engineering, Wuhan University of Technology, Wuhan, China
babamusah94@yahoo.com

Abstract. The utilization of microalgae for the commercial production of clean fuels (biofuel) is considered a suitable renewable energy source compared to fossil fuels. Commercial production of algae is disadvantaged by high capital inputs and low biomass (yield) production threatens the viability of algae biofuel production on large scale. This study, therefore, investigated the suitability of two culture media Blue-Green medium (BG-11) and Bold's Basal medium for the optimum biomass (yield) production of *Chlorella vulgaris* (*C.* vulgaris) and *Scenedesmus meyen (S. meyen)* in a laboratory. The BG-11 medium was more suitable for optimal growth, chlorophyll content development, and biomass production for both *species* as compared to the BB medium. Between the two algae species, *Scenedesmus meyen* produced more biomass 1.3 g (dry weight) in the BB medium as compared to 0.63 g (dry weight) in the BG-11 medium. In contrast, more biomass from C. *vulgaris* was obtained from the BG-11 medium of 0.79 g as compared to 0.04 g in the BB medium at room temperature (25 °C). Given the prospects of algae biofuel production as an alternative renewable energy source, strategies to improve growth performance and yield (biomass) are important to make commercial cultivation of algae cost-effective. This study, therefore, provides an empirical guide for further research aimed at maximizing yield and biomass production by renewable energy industries for clean fuel production.

Keywords: Clean energy · Algae · Culture media

1 Introduction

Over-dependence on fossil fuels has been considered one of the key drivers of global carbon emission which has direct consequences on global warming and exacerbates climate change. Due to the increasing concerns about the impact of fossil fuels on the environment, strenuous efforts are being made for alternative and sustainable clean, and renewable sources of fuel energy to help mitigate the drastic impact of greenhouse gases on the environment. The European Union (EU) recently announced its blueprint for the

X. Chen (Ed.): ICESE 2022, ESE, pp. 144–150, 2023.

attainment of a carbon-neutral target by 2050 under the Paris agreement. The goal is to ensure a significant reduction in carbon emissions by relying more on renewable energy alternatives to mitigate the overall impact of climate change [1]. Algae biofuel has gained some prominence as an alternative green fuel source as compared to crude oil, and coal [2]. Algae-based clean fuel (biofuel) is considered a great potential renewable energy alternative. Many studies have highlighted its feasibility and prospects. However, there are concerns about the cost-effectiveness of commercial cultivation and the production of algae for biofuel utilization. Hence, there is an urgent need to explore innovative strategies to increase and maximize the yield production of algae biomass to make it more sustainable and economical for commercial production. Minimizing the production cost of algae will make it more suitable and commercially viable for large-scale cultivation [3]. Studies showed that a plethora of different biofuel products can be obtained from algae including syngas, lipids, biodiesel, biomethane, biogas, and bioethanol [2]. The great advantage of algae is that many species are capable of surviving in diverse ecosystems such as marine, freshwater, and terrestrial. The ability of algae to produce lipids in their cells just like some oil-producing vegetables makes algae a suitable option for clean fuel production. Studies have reported that over 30% of the cell mass of algae is made up of lipids. Similarly, it is argued that algae can produce about 100 times more oil as compared to any terrestrial plant per acre of land [4].

Algae species are broadly categorized into the following Chlorophyta (green algae), Rhodophyta (red algae), Phaeophyta (brown algae), Xanthophyta (yellow-green algae), Chrysophyta (golden-brown algae), Pyrrophyta (fire algae) and Euglenophyta (Euglenoids). [5]. Similarly, algae could be grouped into two main types based on sizes such as macroalgae and microalgae. Macroalgae are multicellular usually large and visible to the naked eye (e.g. seaweed). Whereas microalgae on the other hand are single-celled algae and microscopic analogous to cyanobacteria (Chloroxybacteria) thus prokaryotic or comparable to green algae (Chlorophyta) thus eukaryotic [6]. Microalgae and cyanobacteria are abundant species in eutrophic ecosystems such as freshwater and they serve a vital role as a major food source for different species of aquatic organisms such as Chironomidae larvae [7]. Microalgae provide a great source of carbon compounds capable of being utilized in various industries including green energy (biofuels) [8]. It is also important to note that microalgae play a very key role in environmental remediation such as carbon dioxide (CO_2) sequestration from the atmosphere and wastewater treatment. In addition to lipids production, other bioproducts including polysaccharides, pigments, proteins, bioactive elements, antioxidants, and vitamins are produced from microalgae for various applications [9]. *Chlorella vulgaris* (*C. vulgaris*) and *Scenedesmus meyen* (*S. meyen*) are some of the most common freshwater green algae species which belong to the phylum of Chlorophyta. It is reported that *C. vulgaris* contains a significant amount of useful nutrients such as fibre (5%), vitamins and minerals (10%), carbohydrate (20%), fat (20%), and protein (45%) w/w dried basis [10]. Even though algae biofuel has been identified as one of the renewable green sources of energy in recent times, the high costs of large-scale production hamper its widespread application [11].

To make the commercial production of algae more viable and cost-effective, there is an urgent need to identify strategies, including improvement in yield production and screening processes for strains (algae species) selection. This necessitates rigorous investigation into the viable cultivation and production of algae biomass to maximize yield for the cost-effective production of biofuel as alternative renewable energy. Although algae production has existed over the past four decades, commercial production of clean fuels and energy from algae sources is yet to be fully understood or developed. Hence, more investigations are required to proffer very flexible cost-effective methods and strategies to help boost yield and biomass production for commercial utilization [12]. This includes the identification of optimum cultivation media, species and nutrient requirements, and culture conditions for effective growth and biomass production. The objectives of this study were *i*) to determine the optimum cultivation medium for *C. vulgaris* and *S. meyen* species, *ii*) to examine the yield of the microalgae species in respect of the two-growth media, and *iii*) to provide empirical evidence on which culture medium is suitable for the two species. The outcome of this investigation is expected to serve as a good guide in the selection of a suitable culture medium and conditions for future production purposes. Hence, a determination of optimum growth medium and yield performance would enhance knowledge and facilitate the optimization of algae production on large scale.

2 Materials and Methods

2.1 Chemicals

Sodium nitrate ($NaNO_3 \geq 99.0\%$), Magnesium sulfate heptahydrate ($MgSO_4.7H_2O \geq 99.0\%$), Sodium chloride ($NaCl \geq 99.5\%$), Dipotassium hydrogen phosphate trihydrate ($K_2HPO_4 \geq 99.0\%$), Potassium hydrogen phosphate ($KH_2PO_4 \geq 99.5\%$), Calcium chloride dihydrate ($CaCl_2.2H_2O$ 99.0 ~ 103.0%), Zinc sulfate heptahydrate ($ZnSO_4.7H_2O \geq 99.5\%$), Manganese (II) chloride tetrahydrate ($MnCl_2.4H_2O \geq 99.0\%$), Molybdenum trioxide (MoO_3, AR 99.5%), Copper (II) sulfate pentahydrate ($CuSO_4.5H_2O \geq 99.0\%$), Cobalt (II) chloride hexahydrate ($CoCl_2.6H_2O \geq 99.06\%$), Boric acid ($H_3BO_3 \geq 99.5\%$), ethylene diamine tetra-acetic acid ($C_{10}H_{16}N_2O_8$, EDTA $\geq 99.5\%$), potassium hydroxide ($KOH \geq 85.0\%$), Iron (II) sulfate heptahydrate ($FeSO_4.7H_2O$, 99.0 ~ 101.0%) and Sulfuric acid (H_2SO_4) conc., Sodium hydroxide ($NaOH \geq 96.0\%$), Ethanol absolute ($C_2H_6O \geq 99.7\%$), Potassium bromide ($KBr \geq 99.0\%$), Hydrochloric acid ($HCl \geq 98\%$), Phosphoric acid ($H_3PO_4 \geq 85.0\%$), Nitric acid (HNO_3 65.0 ~ 68.0%), Citric acid ($C_6H_8O_7.H_2O \geq 99.5\%$), Ethylenediamine tetra-acetic acid disodium salt dihydrate ($C_{10}H_{14}N_2Na_2O_8.2H_2O$, $EDTANa_2 \geq 99.0\%$), Sodium carbonate ($Na_2CO_3 \geq 99.8\%$), Ammonium ferric citrate (AR, Fe:20.5–22.5%), Sodium molybdate dihydrate ($Na_2MoO_4.2H_2O$, 98%), Sodium acetate trihydrate ($C_2H_3NaO_2.\ 3H_2O$, $\geq 99.0\%$). All chemicals and reagents were supplied by Sinopharm Chemical Reagents, China except MoO_3, $Na_2MoO_4.2H_2O$, and Ammonium ferric citrate, which was supplied by Aladdin®. All chemicals and reagents were of analytical grades.

2.2 Preparation of Growth Media

Pure culture strains of *Chlorella vulgaris* (*C. vulgaris*) and *Scenedesmus meyen (S. meyen)* were purchased from the Institute of Hydrobiology, Chinese Academy of Sciences, Wuhan. The strains were sub-cultured in the laboratory in sterilize Bold's Basal

Medium (BBM) and Blue-Green medium (BG11). During the preparation of each stock solution for the BB medium, the following masses (g) were dissolved in volumetric flasks containing 400 mL each of ultra-pure water (Resistivity 18.25 M_{Ω} cm): $NaNO_3$ (10.0 g), $MgSO_4.7H_2O$ (3.0 g), NaCl (1.0 g), K_2HPO_4 (3.0 g), KH_2PO_4 (7.0 g) and $CaCl_2.2H_2O$ (1.0 g). The trace elements solutions were prepared by dissolving the following masses in a litre (L) of acidified ultra-pure water: $ZnSO_4.7H_2O$ (8.82 g), $MnCl_2.4H_2O$ (1.44 g), MoO_3 (0.71 g), $CuSO_4.5H_2O$ (1.57 g), and $CoCl_2.6H_2O$ (0.49 g). The iron (Fe) solution was prepared by dissolving 4.98 g of $FeSO_4.7H_2O$ in 1 L of acidified ultra-pure water. This was achieved by the addition of 1 mL concentrated sulfuric acid to make up 1 L solution. The boron solution was prepared by dissolving 11.42 g of H_3BO_3 in 1 L of ultra-pure water. Additionally, the ethylene diamine tetra-acetic acid (EDTA) solution was prepared by dissolving 50 g of EDTA and 31 g of KOH in a litre of ultra-pure water. In the BG11 medium, the following stocks were prepared in 500 mL volumetric flasks containing the appropriate solutes using ultra-pure water as solvent. $NaNO_3$ (75.0 g), K_2HPO_4 (2.0 g), $MgSO_4.7H_2O$ (3.75 g), $CaCl_2.2H_2O$ (1.80 g), $C_6H_8O_7$ (0.30 g), Ammonium ferric citrate (0.30 g), $EDTANa_2$ (0.05 g) and Na_2CO_3 (1.0 g). The composition of the trace elements per liter was H_3BO_3 (2.86 g), $MnCl_2.4H_2O$ (1.81 g), $ZnSO_4.7H_2O$ (0.22 g), $Na_2MoO_4.2H_2O$ (0.39 g), $CuSO_4.5H_2O$ (0.08 g) and $CoCl_2.6H_2O$ (0.05 g). For the preparation of the culture media, the pH was adjusted to 7.1 using 0.1 M NaOH or HNO_3.

2.3 Sterilization of Growth Media and Cultivation

The sterilization of the culture media was conducted at a temperature of 121 °C for 20 min (min). The sterilization was carried out in a vertical heating pressure steam sterilizer (LDZX-50KBS) manufactured by Shanghai Shenan Medical Instrument Factory (SSMIF), China. The screwcaps of the bottles that contained the culture media were not fully tightened up to enable the escape of heated air after the culture media were loaded in the autoclave. After the sterilization process, the autoclave was allowed 160 min to cool down until the chamber temperature reached 80 °C or below before the loads were safely taken out from the sterilizer. The sterilized culture media were allowed to cool down at room temperature (25 °C) for subsequent use. All the Borex flasks used for the culture media were capped with cotton stoppers before being sterilized under the same conditions as indicated previously. The *C. vulgaris* and *S. meyen* inoculum were grown in two 1 L Borex flasks containing the various culture media for 21 days at a temperature of 25 ± 2 °C in an intelligent light incubator. The incubator (GXZ-380B) has a temperature control range of 0 to 50 ^{O}C with maximum illumination of 15000 lx. The illumination employed was in the range of 400 t 500 lx. The incubator was manufactured by Ningbo Jiangnan Instrument Co. Ltd, Ningbo, China. The illumination was alternated such that there was 12 h illumination and 12 h darkness during the 21 days. The culture media were regularly renewed as required to ensure the continuous supply of adequate nutrients for metabolic activities. The cultures were aerated (oxygen enrichment) with a constant flow rate of 200 l/s. The gas flow metres helped in regulating the amount of air that entered the culture at any given time for the optimization of the air bubbles. The air injected into the inoculated culture is very critical because it prevents cell sedimentation and agglomeration at the bottom of the flasks. Furthermore, the concentration of nutrients at the bottom is minimized, and it also increases the exposure of light to the cells at the bottom of the flasks [13].

3 Microalgae Harvesting and Yield

After 21 days of cultivation, the two microalgae sample species were allowed 24 h for adequate settling down of the biomass without providing any aeration. The algae slurry was dewatered after harvesting through centrifugation at a speed of 4000 g for 15 min at a temperature of 20 ^{O}C using Centrifuge 5810 R (Eppendorf, Germany). The biomass was then washed several times with ultra-pure water until a neutral pH (7) was attained to ensure the complete removal of any impurities and oven-dried at 105 °C for 24 h. The biomass was grounded in a mortar with a pestle and weighed. The various processes of cultivation and harvesting have been outlined in Fig. 1. Based on the two-culture media, *S. meyen* produced more biomass in both BB medium and BG 11 medium as compared to *C. vulgaris*. Whereas, with the BB medium, the *S. meyen* produced biomass of 0.63 g (dry weight), compared to the 1.3 g (dry weight) biomass obtained using the BG 11 medium for the same species. For the *C. vulgaris,* the biomass obtained from the BB medium was 0.04 g while 0.79 g (dry weight) biomass was obtained from the BG 11 medium at a temperature of 25 °C. Similar growth conditions and yield performance for *C. vulgaris* in BG 11 medium were reported in comparison to BB medium [14]. A similar comparative investigation conducted by Bawja et. Al., (2017) revealed that the BG-11 medium is the best culture medium for high biomass yield for *Chlorella pyrenoidosa*, *Chlorococcum acquaticum,* and *Scenedesmus obliquus* [15]. The production efficiency of microalgae can be determined via many parameters such as biomass, lipids, and carbohydrates [16]. Microalgae serve as promising alternative biomaterials for the effective sequestration of toxicants in the environment via biosorption and bioaccumulation techniques due to their physical and chemical properties [17] BG 11 medium is rich in N as compared to BB medium; this may have influenced the rate of biomass production as observed in this

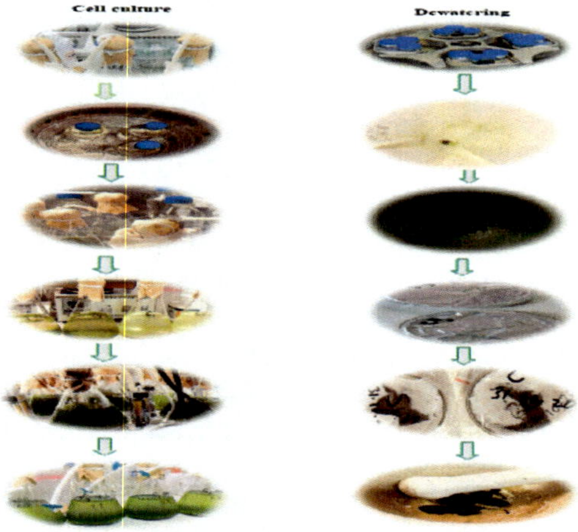

Fig. 1. Schematic experimental algae culture and dewatering processes in the laboratory.

investigation. It is reported that *C. vulgaris* grows well in high nutrient concentrations and its growth is inhibited by an unbalanced N:P ratio and inadequate P [14].

4 Conclusion

Given that the high cost of algae biomass production especially in commercial quantity is one of the key setbacks that hinders biofuel and clean energy, this work sought to provide clear evidence and guidance on the appropriate selection of culture medium to maximize yield and biomass production. In this study, growth and biomass (yield) production have been investigated using *Chlorella vulgaris* and *Scenedesmus meyen* in the laboratory. The two algae species were cultured in Blue-Green (BG-11) and Bold's Basal (BB) medium respectively. The BG-11 was more suitable for both growth and biomass production for the two algae species as compared to the Bold's Basal (BB) medium. The biomass (dry weight) of *Scenedesmus meyen* was higher in the BG-11 and BB medium, unlike *Chlorella vulgaris*. This is indicative that the BG-11culture medium is more suitable for boosting biomass yield production under optimized laboratory conditions. It is anticipated that this would serve as empirical evidence to guide further research into strategies to maximize algae biomass for commercial production.

Acknowledgement. The National Social Science Fund of China: The Restructure of International Economic Regulations (19ZDA054).

References

1. Potrč, S., Čuček, L., Martin, M., Kravanja, Z.: Sustainable renewable energy supply networks optimization – the gradual transition to a renewable energy system within the European Union by 2050. Renew. Sustain. Energy Rev. **146** 111186 (2021)
2. Hossain, J.., Jahan, R..: Biofuel: marine biotechnology securing alternative sources of renewable energy. In: Maddela, N.R., GarcíaCruzatty, L.C., Chakraborty, S. (eds.) Advances in the Domain of Environmental Biotechnology. EMB, pp. 161–194. Springer, Singapore (2021). https://doi.org/10.1007/978-981-15-8999-7_7
3. El-Sayed, W.M.M., Ibrahim, H.A.H., Abdrabo, M.A.A., Abdul-Raouf, U.M.: Algal-based biofuel: challenges and future perspectives. In: Biofuel from Microbes and Plants, pp. 23–42 (2021)
4. Singh, A., Nigam, P.S., Murphy, J.D.: Mechanism and challenges in commercialisation of algal biofuels. Biores. Technol. **102**, 26–34 (2011)
5. Bailey, R.: 7 Major Types of Algae ThoughtCo 8 (2020)
6. Khan, M.I., Shin, J.H., Kim, J.D.: The promising future of microalgae: current status, challenges, and optimization of a sustainable and renewable industry for biofuels, feed, and other products. Microbial Cell Factories **17**, 1–21 (2018)
7. Stanković, N., et al.: Can phytoplankton blooming be harmful to benthic organisms? The toxic influence of Anabaena sp. and Chlorella sp. on Chironomus riparius larvae. Sci. Total Environ. **729**, 9 (2020)
8. Das, P., Aziz, S.S., Obbard, J.P.: Two phase microalgae growth in the open system for enhanced lipid productivity. Renew. Energy **36**, 2524–2528 (2011)

9. Brennan, L., Owende, P.: Biofuels from microalgae-A review of technologies for production, processing, and extractions of biofuels and co-products. Renew. Sustain. Energy Rev. **14**, 557–77 (2010)
10. Belasco, W.: Algae burgers for a hungry world? The rise and fall of chlorella cuisine. Technol. Cult. **38**, 608–634 (1997)
11. Loftus, S.E., Johnson, Z.I.: Cross-study analysis of factors affecting algae cultivation in recycled medium for biofuel production. Algal Res. **24**, 154–166 (2017)
12. Efroymson, R.A., Jager, H.I., Mandal, S., Parish, E.S., Mathews, T.J.: Better management practices for environmentally sustainable production of microalgae and algal biofuels. J. Clean. Prod. **289**, 125150 (2021)
13. Apel, A.C., Weuster-Botz, D.: Engineering solutions for open microalgae mass cultivation and realistic indoor simulation of outdoor environments. Bioprocess Biosyst. Eng. **38**, 995–1008 (2015)
14. Munir, N., Imtiaz, A., Sharif, N., Naz, S.: Optimization of growth conditions of different algal strains and determination of their lipid contents. J. Animal Plant Sci. **25**, 546–553 (2015)
15. Bajwa, K., Bishnoi, N.R., Kirrolia, A., Sharma, J., Gupta, S.: Comparison of various growth media composition for physio-biochemical parameters of biodiesel producing microalgal species (Chlorococcum aquaticum, Scenedesmus obliquus, Nannochloropsis oculata and Chlorella pyrenoidosa). Eur. J. Biotechnol. Biosci. **5**, 27–31
16. Ahmad, A., Bhat, A.H., Buang, A., Shah, S.M.U., Afzal, M.: Biotechnological application of microalgae for integrated palm oil mill effluent (POME) remediation: a review. Int. J. Environ. Sci. Technol. **16**, 1763–88 (2019)
17. Mantzorou, A., Navakoudis, E., Paschalidis, K., Ververidis, F.: Microalgae: a potential tool for remediating aquatic environments from toxic metals. Int. J. Environ. Sci. Technol. **15**, 1815–1830 (2018)

Hardware in the Loop and Experiment Validation for a PV Inverter Under Flexible Power Control

Achara Pichetjamroen[1](✉), Nithiphat Teerakawanich[1], Phacharawat Chindamanee[1], and Natthawuth Somakettarin[2](✉)

[1] Department of Electrical Engineering, Faculty of Engineering, Kasetsart University, Bangkok, Thailand
achara.pi@ku.ac.th

[2] Department of Electrical Engineering, Rajamangala University of Technology Thanyaburi, Patumthani, Thailand
natthawuth.s@en.rmutt.ac.th

Abstract. Solar intermittency is the primary problem for the PV power generating system with a principal challenge of PV of the weather. Flexible power point tracking (FPPT) control can be incorporated into the system to allow dual-mode control approach between Power Ramp-Rate Control (PRRC) and Maximum Power Point Tracking (MPPT) control to address these issues. This paper develops a testbed platform which includes a hardware-in-the-loop (HIL) simulator, a real inverter, and loads to validate the operation of FPPT control under the realistic irradiance profile. The scaled down setup of the proposed framework with the inverter output limit of 600 W is used for the experimental validation. The output power from the HIL inverter model and the real inverter are compared in 4 modes of operation. Overall, the results from the HIL simulation model agree well with the experiment with slight deviation in the ramp down mode. The experimental framework presented here is intended for stand-alone PV system test, however, we should be able to extend for a test of higher power and more complex system such as microgrid where inverters will be grid connected and loads will be of various power profile.

Keywords: PV inverter · Flexible power point tracking · Hardware in the loop · Solar irradiance fluctuation

1 Introduction

Solar intermittency is the primary problem for the PV power generating system with a principal challenge of PV of the weather. Considering the dynamic nature of the PV production system, passing clouds can pose an additional difficulty to the system by the high-power ramp rate and power fluctuations. The abrupt variations in PV power can cause fluctuations in voltage and frequency as well as a decrease in system stability [1–3]. Flexible power point tracking (FPPT) control can be incorporated into the system to allow dual-mode control approach between Power Ramp-Rate Control (PRRC) and Maximum

X. Chen (Ed.): ICESE 2022, ESE, pp. 151–157, 2023.

Power Point Tracking (MPPT) control to address these issues [4, 5]. The additional benefit of FPPT control is that it can limit or reduce the size of the needed energy storage for the resilience of a PV system which has been demonstrated by simulation in [6].

This paper develops a testbed platform which includes a hardware-in-the-loop (HIL) simulator, a real inverter, and loads to validate the operation of FPPT control under the realistic irradiance profile. The proposed ramp rate control algorithm is implemented and validated by both HIL simulation and experiment. Details of the developed test framework are discussed in the next section.

2 HIL Simulation and Experiment Framework

The configuration of the HIL and hardware testbed of a PV power system is illustrated in Fig. 1. In this framework, the system consists of 2 parts: a HIL model and an experimental hardware. The simulated system includes PV panels connected to a DC-DC boost converter under FPPT control for stepping up the supplied voltage from PV panels, and a single-phase inverter unit for converting Dc voltage into ac for load utilization. The ramp-rate limit is given to the FPPT algorithm (Dual-Mode) to provide an appropriate PWM signal for the boost converter. The dual-mode was introduced to control the overall system to obtain maximum power point, and also the ramp rate is limited.

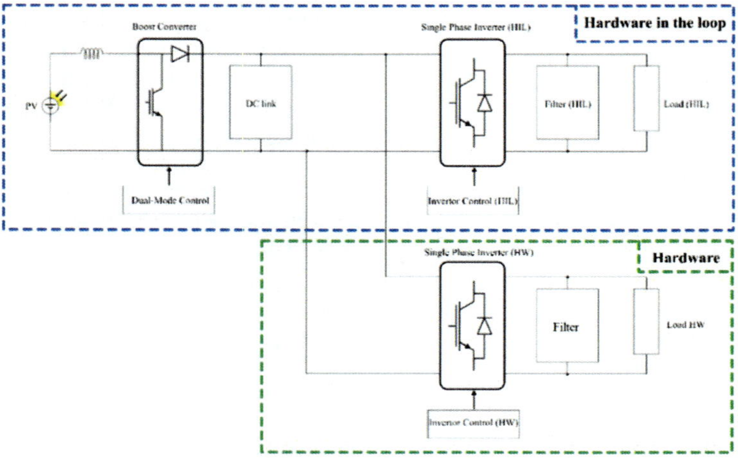

Fig. 1. Circuit configuration of PV power system hardware in the loop (HIL) and hardware.

The ramp rate can be determined by (1). Changes in PV power are limited to 10% of the installed PV power per minute [4], but this ramp rate (R_r^*) can be modified to different values as necessary.. The objective function in PRRC ramp-up and ramp-down control units is used to activate these units (2)

$$R_r(t) = \frac{P_{pv}(t) - P_{pv}(t - \Delta t)}{\Delta t} \quad (1)$$

$$P_{pv}^{*} = \begin{cases} P_{MPPT}, & \text{when } R_r(t) \le R_r^{*} \\ P_{limitramp-up}, & \text{when } R_r(t) > R_r^{*} \\ P_{limitramp-down}, & \text{when } R_r(t) > R_r^{*} \end{cases} \quad (2)$$

The simulation parameters refer to the previous simulation system in [1] and [2]. In the hardware framework, a simulated DC link voltage level is signaled out to a programmable DC power supply as a source to an inverter of the experiment system. PWM signals of both the HIL model inverter and the hardware inverter are generated within the real-time HIL unit. The output voltage of the hardware inverter supplies power to an AC electronic load, which is controlled to consume with the same load profile as the HIL load model. Therefore, both inverters are under the same operating conditions. The parameters of the inverter are listed in Table I.

Table 1. Experimental parameters of the inverter.

System parameters	Values
LCL filter	$L_1 = 1.5$ mH, L_2 0.5 mH, $C_f = 1.0$ µF
Switching frequency for inverter	$f_{inv} = 20$ kHz
Nominal DC-link voltage	$V_{dc}^{*} = 200$ V
Output voltage (RMS)	$V_g = 110$ V
Output frequency	$\omega_0 = 377$ rad/s
Max. Load power	600 W

The experimental setup in Fig. 2 composes of a programmable DC power supply which emulates DC link voltages from the HIL simulation, an AC electronic load.

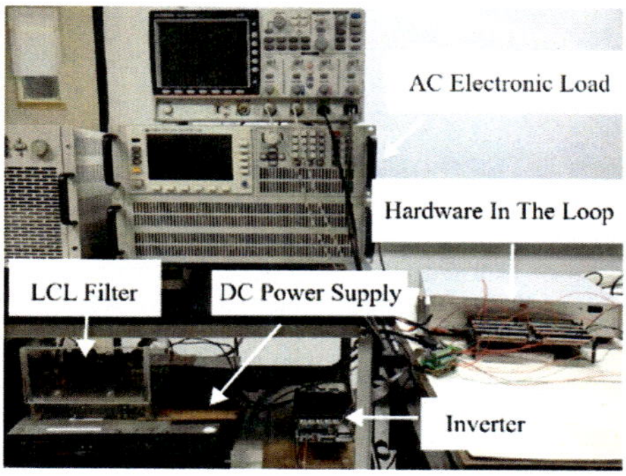

Fig. 2. Experimental setup of the single phase inverter connected load.

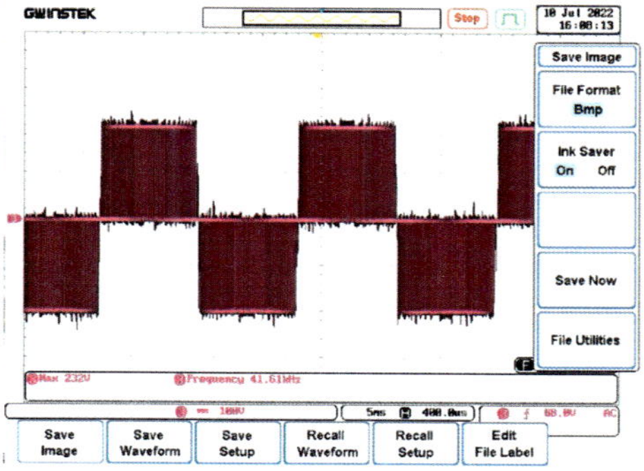

Fig. 3. Inverter output waveform before connecting to LCL filter.

whose load power is controlled by the HIL unit. The inverter is a SiC 2-level inverter switching at 20 kHz with PWM signals from the HIL unit. The inverter output voltage is regulated at 110 V_{rms} and 60 Hz as shown in Figs. 3 and 4.

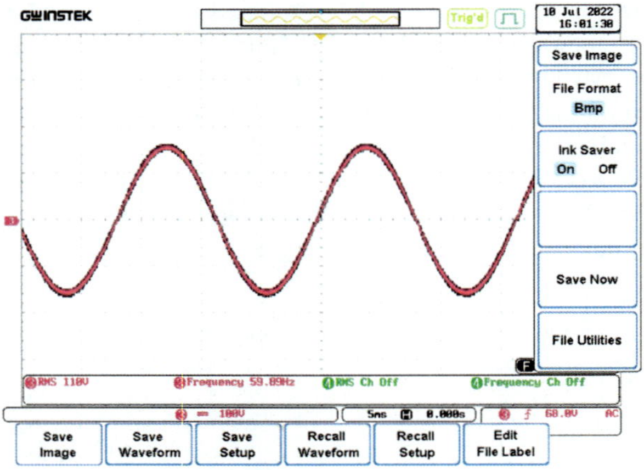

Fig. 4. Voltage output waveform after LCL filter

3 HIL Simulation and Experimental Result

The HIL experimental framework that we developed is applied to test the PV inverter FPPT operation under realistic solar irradiance profile. The scaled down setup of the proposed framework with the inverter output limit of 600 W is used for the experimental validation. The testing solar irradiance profile composes of 4 main features as demonstrated in Fig. 5. The output power from the HIL inverter model and the real inverter are compared and also shown in Fig. 5. Figure 6 is illustrated the ramp rate and DC link

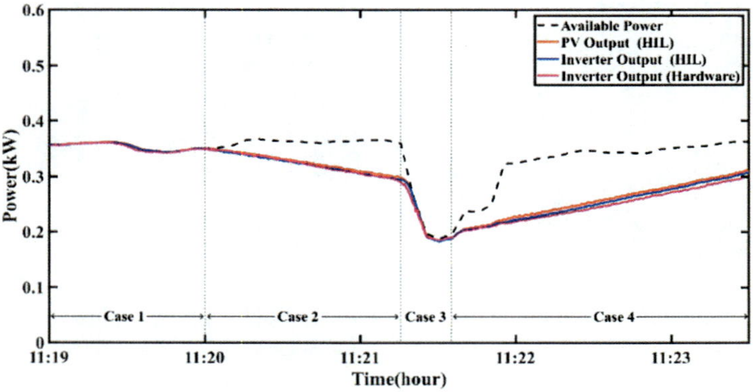

Fig. 5. PV power generation profiles under various situations.

voltage that can maintained until the end of this period where a sudden change of solar irradiance really occurs.Voltage and current waveforms of the output to the load are set at unity power factor as shown in Fig. 7. The results are as follows.

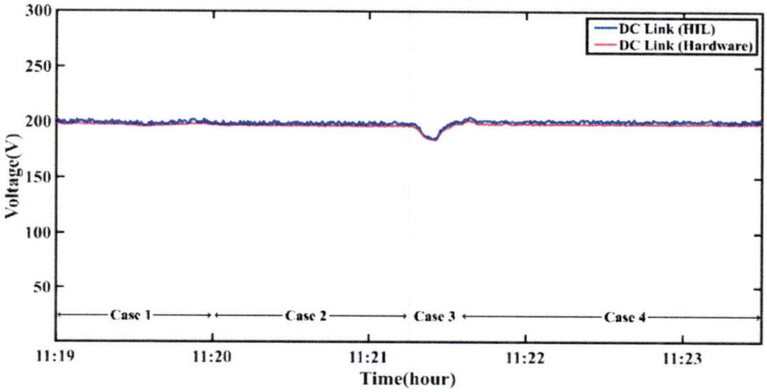

Fig. 6. DC link voltage under various scenarios.

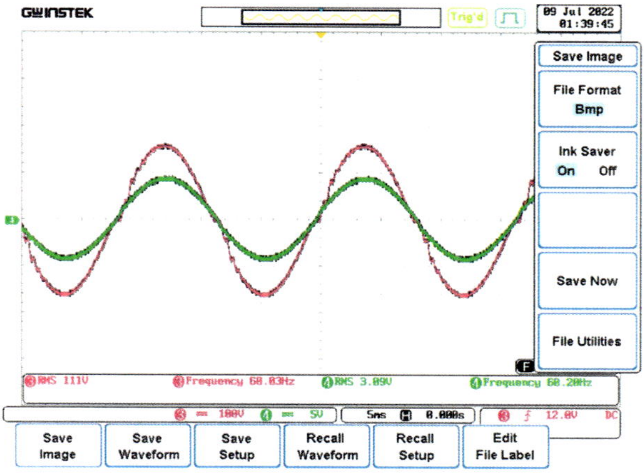

Fig. 7. Voltage and current waveforms at load.

3.1 Case 1: Inverter Under Normal MPPT Operation

The first part of the solar profile exhibits a slight rate of change where MPPT control is sufficient to track the available output power. Both HIL inverter and the hardware inverter can provide the maximum power with accurate tracking. DC link voltage is kept constant during this period. Both HIL model and the hardware inverter give the exact behavior.

3.2 Case 2: Inverter Under FPPT Ramp Down Operation

The second part of the solar profile includes a part where the trigger from the cloud coverage prediction system arises 1–2 min in advance [7]. The trigger signals the inverter at 11:20 to operate in FPPT mode to reduce the output power down in advance to avoid a huge drop of the output that may cause unwanted dynamic. We can notice a gradually decrease output power with predefined ramp rate from the inverter (60 W/min in this experiment). Only a slight difference between HIL simulation model and the experiment can be observed. The output from the experimental inverter during the ramp down control shows approximately the same ramp rate of 60 W/min comparing to the HIL model.

3.3 Case 3: Inverter under Natural Ramp Down

The inverter in this mode encounters a sudden drop in PV power due to the weather as predicted beforehand. The output ramp rate cannot be controlled without energy storage [8] and the output of the inverter will drop sharply as a result. Therefore, dynamic of the system from power changes will be unavoidable in this situation as can be seen from the DC link voltage. At 11:21:25, the DC link voltage dip of 8% is observed. However, this dynamic is not as critical as it would have been in case no ramp down control. The abrupt power is limited to only 115 W whereas it would be 180 W without ramp down action (36% difference). We could imply the advantage of FPPT in this situation where ramp down control is required.

3.4 Case 4: Inverter Under FPPT Ramp up Operation

The inverter in this mode yields an output power at the predefined ramp rate to restrict the change of PV power. Both the results from the HIL model and the experiment exhibits equivalent behavior in this mode. With gradually increase in output power, DC link voltage is maintained almost constant during the ramp up operation. FPPT is normally easy to implement in this operation and energy storage is not required. Overall, the results from the HIL simulation model agree well with the experiment with slight deviation in the ramp up mode due to the accuracy of the AC electronic load at low power.

4 Conclusion

The operation of FPPT control of the PV inverter has been validated both in real-time HIL simulation and with the real inverter hardware. The control algorithm has been tested with the real solar irradiance profile. Using HIL has an advantage of reproduction results. However, any control algorithm still requires hardware demonstration to validate its benefit in practical implementation. Therefore, we propose an experiment framework to compare the results from real-time HIL simulation and the real inverter and load. The experimental framework presented here is intended for testing a stand-alone PV system, however, we should be able to extend for a test of higher power and more complex system such as microgrid where inverters will be grid connected and loads will be of various power profile.

References

1. Chindamanee, P., Pichetjamroen, A., Teerakawanich, N.: Dual-mode operation of pv power converters for flexible power tracking. In: 2021 9th International Electrical Engineering Congress (iEECON), pp. 133–136. IEEE (2021)
2. Pichetjamroen, A., et al.: Effect of flexible power control on batteries requirement and performance in PV power system. Energy Rep. **8**, 338–344 (2022)
3. Pichetjamroen, A., et al.: A Study on performances of flexible power control with empirical lithium-ion battery modeling in PV power systems. In: 2022 19th International Conference on Electrical Engineering/Electronics, Computer, Telecommunications and Information Technology (ECTI-CON). IEEE (2022)
4. Sangwongwanich, A., Yang, Y., Blaabjerg, F.: A cost-effective power ramp-rate control strategy for single-phase two-stage grid-connected photovoltaic systems. In: 2016 IEEE Energy Conversion Congress and Exposition (ECCE). IEEE (2016)
5. Sukumar, S., et al.: Ramp-rate control smoothing methods to control output power fluctuations from solar photovoltaic (PV) sources—a review. J. Energy Storage **20**, 218–229 (2018)
6. Atif, A., Khalid, M.: Saviztky-Golay filtering for solar power smoothing and ramp rate reduction based on controlled battery energy storage. IEEE Access **8**, 33806–33817 (2020)
7. Teerakawanich, N., Leelaruji, T., Pichetjamroen, A.: Short term prediction of sun coverage using optical flow with GoogLeNet. Energy Rep. **6**, 526–531 (2020)
8. Dickeson, G., et al.: Ramp rate control for PV plant integration: experience from Karratha Airport's hybrid power station. In: 36th European Photovoltaic Solar Energy Conference and Exhibition (2019)

Bioclimatic Architecture as a Design Basis for the Use of Renewable Energies and Sustainable Development, the Case of Sustainable Social Housing in Patagonia, Aysén, Chile

Samuel Carmona[1](✉), Stefania Pareti[2], Loreto Rudolph[1], and David Flores[3]

[1] Universidad Técnica Federico Santa María, Valparaíso, Chile
{samuel.carmona,loreto.rudolph}@sansano.usm.cl
[2] Universidad Andrés Bello, Santiago, Chile
s.pareti@uandresbello.edu
[3] Pontificia Universidad Católica de Chile, Santiago, Chile
daflores2@uc.cl

Abstract. The objective of this study is to explore sustainable prototypes of social housing in Chile, as the basis of a bioclimatic architectural design. This focuses on the use of renewable energy and sustainable development. Regarding the case, it is about social housing prototypes that acquire bioclimatic characteristics through climatic conditioning strategies, therefore they are related to energy efficiency, the place in which they are located and sustainable construction with low CO2 emissions. This case is selected, because the construction industry today presents a constant challenge for sustainable development, therefore these sustainable housing proposals appear as an applicable solution for this issue. The methodology is carried out through a constructive analysis of a typical dwelling to understand the decisions made and how bioclimatic strategies influence the habitability of the building. This will be complemented with the study of secondary sources to explore the application of the referenced strategies. It is concluded that bioclimatic social housing in Aysén allows sustainable development over time, given the strategies with which they are designed, the material chosen and their efficiency with respect to energy saving.

Keywords: Social housing · Sustainability · Renewable energy · Bioclimatic architecture

1 Introduction

The Aysén Region is Located in the Southern Zone of Chile (Fig. 1), Being the Gateway to Chilean Patagonia. Its Main Feature is Its Geography, Resulting from the Intense Glacial Activity in the Area that Shaped Its Landscape, Which Still Has Glaciers in Certain Sectors Known as North and South Ice Fields [1].

X. Chen (Ed.): ICESE 2022, ESE, pp. 158–163, 2023.

The Region Consists of Two Types of Climates. On the One Hand, the Archipelagos and Fjords Are Characterized by a Rainy Oceanic Climate. On the Other Hand, the Eastern Sector of the Region Has a Steppe Climate, Cold and Dry.

In This Geographical and Climatic Context, the "SUstainable Social Housing in Patagonia" Initiative is Carried Out, Which Seeks to Implement a Prototype of Sustainable Social Housing with Wood as the Main Material. A Bioclimatic Architectural Approach Would Be Used, Seeking to Reduce Heating Costs and Reduce CO2 Emissions Resulting from the Burning of Firewood, Which is Currently the Main Fuel in the Area When It Comes to Heating Homes.

Fig. 1. Map of chile, Aysén region. Own elaboration.

The project is inserted within a global context that demands more and more sustainable architectural solutions that are in tune with the new needs that climate change implies. In this sense, the study takes the following concepts as central themes to be carried out: (1) Sustainable development (2) bioclimatic architecture (3) energy efficiency in housing (4) sustainable housing.

Sustainability is development that meets the needs of the present without compromising the capacity of future generations, without compromising the economy, care for the environment and social welfare. The promotion of social development seeking cohesion between communities and cultures to achieve satisfactory levels of quality of life [3]. The concept of sustainable development is not only focused on environmental protection, rather sustainable development also implies economic growth that provides justice and opportunities for all the peoples of the world and not just a minority, stopping destroying natural resources and global carrying capacity. Sustainable development encompasses the economic, ecological and social framework.

Prior to the use of fossil fuels, traditional buildings paid significant attention to local climatic elements, unlike modern means of heating or air conditioning. After the recent energy crisis, there is a new interest in techniques that save energy and, in particular, in natural techniques [4]. Bioclimatic architecture manages to incorporate, from the first stages of design, strategies and resources that allow taking advantage of the favorable conditions of the climate and the natural environment [5], allowing the generation of both comfortable interior conditions and collaborating with the minimization of the energy impact of the building. Bioclimatic design manages to optimize resources through morphology, orientation, materials, color configuration and other design variables.

One of the main sources of energy expenditure in the home is air conditioning, which in both winter and summer is required to heat or condition the environment, in addition to lighting and the use of electrical appliances. Accordingly, buildings can be classified as more or less energy efficient depending on thermal insulation and the use of renewable energies [6]. Every building must be labeled with the Energy Label, documentation that qualifies the energy demand of each construction on an A-G scale, the first letters being those with a lower energy demand, and the last ones with a higher energy demand. The social housing system in Patagonia combines a series of passive measures to be energy efficient, both capturing and retaining energy. Either by capturing direct solar radiation, associated with thermal mass and optimization of the heat generated by the wood stove. Sustainable homes both in Chile and in other countries of the world are being increasingly valued by architects and also owners who seek to contribute to caring for the environment through homes that not only have good design but also materials. That are used to help to contribute to the reduction of contamination [7].

2 Methodology

The investigation will be carried out through a study of secondary sources associated with the case, mainly public information provided by the office of architects responsible for the project, where it is characterized from an architectural and bioclimatic perspective, to establish its role as an object of study and the intention of understand the architecture associated with the place and its advantages over sustainability. Said information was complemented with texts and theoretical discourses that move within the fields of sustainable and bioclimatic architecture.

Secondly, a constructive analysis was carried out, proposing the main operations that make up this type of housing. This exercise allows recognizing the motivations behind the decisions made, both structural and material, to propose them as air conditioning strategies with respect to the environmental conditions of the location.

3 Discussion

In Chile, with respect to social housing, it is not possible to satisfy all the necessary requirements due to insufficient resources; being able to meet only the minimum conditions estimated as acceptable, and therefore, there is an urgent need to make it evolve over time, tending to its transformation and improvement [8].

As a consequence of the above, sustainable housing development plans are beginning to be implemented that have a higher quality standard, whose main objective is to achieve high energy efficiency and environmental quality [9].

Buildings are responsible for a high proportion of global energy consumption, hence the importance of this theme, in which all opportunities to apply sustainable strategies are evaluated to minimize not only environmental impact problems, but also to do more efficient energy consumption in buildings [10].

The activity related to the construction of buildings and communities is a critical factor in the degradation of the environment, given the high energy consumption and CO2 emissions it presents [11]. It is estimated that 26% of the country's total energy is destined only for the operation of buildings [12]. Furthermore, the built stock contributes 33% to the emission of greenhouse gasses and particulate matter during its construction and operation [13].

Applying bioclimatic design in buildings also helps preserve the environment, integrating man into a more balanced ecosystem through spaces, generating the least possible impact [14].

With respect to the case, it is a bioclimatic prototype of sustainable social housing in wood in the city of Aysén, which in material terms has a modular post and beam construction system, to allow versatility in the use of interior spaces and drastically reduce the footprint of carbon in construction due to the use of wood.

The house is made up of walls with high thermal inertia, these are fed by solar radiation that falls through the translucent areas of the walls and the roof [15], materially they are made up of a double layer of alveolar polycarbonate, separated by a chamber unventilated air. This causes the nucleus to function as a greenhouse that allows clothes and firewood to dry, reducing the contamination of the latter. This greenhouse also functions as a thermal buffer space that raises temperatures in the winter [15].

The orientation of the house is always given by the north orientation, privileging the lighting in search of the greatest solar radiation. Regarding the central wall, in addition to direct solar radiation energy, it is fed by the heat generated by the wood stove proposed in the project.

Due to the adaptation to the climate and the existence of constant rainfall throughout the year, the central nucleus has a rainwater accumulation system through a cistern, which is surrounded by a serpentine in the wall associated with the wood stove, allowing hot water through the use of biomass, a 100% renewable energy source.

The proposed housing system combines a series of passive measures for capturing and retaining energy, among which are the capture of direct solar radiation associated with thermal mass, insulation levels well above what is required by the thermal standard, and optimization of the use of the heat generated by the wood stove [15] (Fig. 2), due to this there is a great reduction in the annual energy consumption for heating.

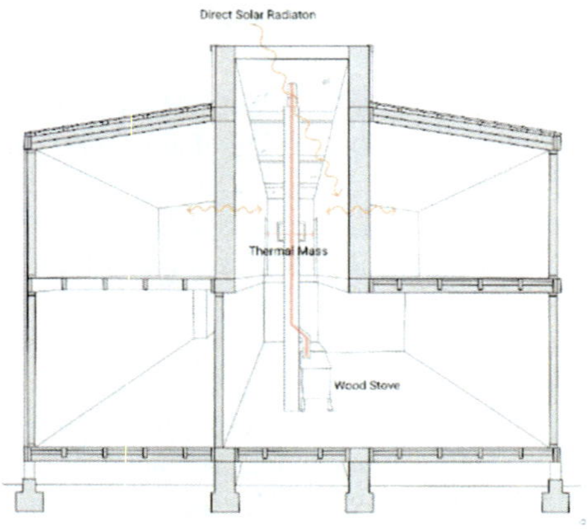

Fig. 2. Structural scheme and passive air conditioning strategies. Own elaboration.

4 Conclusion

Social housing in Chile has a series of deficiencies, establishing its habitability standards as minimum. This is complemented by the low purchasing power of the beneficiary families, giving a generally negative result when understanding the context of maintenance and conditioning necessary for a home. Understanding this problem, the case study is presented as an alternative to the difficulties involved in building social housing in the country.

Before highlighting the construction specificities, it is important to highlight how quality habitability conditions can be established from the design process without necessarily appealing to external factors that ultimately imply higher costs, this in reference to the implementation of passive thermal conditioning strategies in a region characterized by low temperatures. Thus, the importance of the process for the consolidation of sustainable constructions is highlighted.

In this line, the importance of wood compared to other construction materials such as concrete stands out when understanding the impact of the industry within the current problem of global warming. Thus, the carbon footprint of wood is clearly lower than that of other materials, added to the concept of working with local materials and specific design decisions, such as the construction of walls with high thermal inertia or working with air chambers. to improve thermal conditions and reduce the need for external agents, a significant reduction in the impact of housing on the environment is achieved.

This is complemented with the understanding of the geographical area and the customs of the target audience, highlighting the role of the wood-burning stove as a central element that, by working with the thermal inertia walls, allows the complete heating of the home; this considering that the use of this type of stoves is part of the daily life in this sector given the low temperatures, becoming an articulating element of the house, where the daily chores are carried out.

Considering this, it is concluded that bioclimatic social housing in Aysén allows sustainable development, given the strategies with which they are designed, their materiality and the efficiency exposed with respect to the energy savings they achieve.

References

1. Ortega, H., Brüning, A.: Toponimias selectas. Aisén: Panorama histórico y cultural de la XI Región. LOM (2004)
2. Álamos, N., et al.: Caracterización del mercado de la leña en Chile y sus barreras para la transición energética (2020)
3. Robles, C., García, J.: Desarrollo sostenible. realidad o ficción, ábaco, 2 época, no. 17/18, pp. 58–71. Centro de Iniciativas Culturales y Estudios Económicos y Sociales (CICEES) (1998)
4. Camous, R., Watson, D.: L'habitat bioclimatique: De la conception à la construction. L'Etincelle (1979)
5. Maiztegui, B.: Arquitectura bioclimática en latinoamérica: Técnicas naturales para economizar energía. Plataforma Arquitectura (2021). Accessed 17 Apr 2022. https://www.plataformaarquitectura.cl/cl/956847/arquitectura-bioclimatica-en-latinoamerica-tecnicas-naturales-para-economizar-energia
6. Asdrubali, F., Desideri, U.: Handbook of energy efficiency in buildings. building envelope (2019)
7. Haeseong, J.: The new challenge of urban and architectural design for sustainable development. Int. J. Sustain. Build. Technol. Urban Develop. **4** (2013)
8. Haramoto, E.: Vivienda social: un desafío para la sustentabilidad del desarrollo. Revista INVI **10**(24), 18–33 (1995)
9. Roldan, J., Diaz, B., Andrade, P., Castro, C., Gutiérrez, C., Uribe, S.: Bioclimatic sustainable group housing in Santiago de Chile: proposal for an interior growth model. Avances en Energías Renovables y Medio Ambiente, vol. 12 (2008)
10. Guerra, M.: Arquitectura Bioclimática como parte fundamental para el ahorro de energía en edificaciones. Revista Ing-Novación **5**, 123–133 (2013)
11. Blanco, J.: Vivienda Social, Regeneración y Sustentabilidad: Hacia la Ecorehabilitación (Eco-Retrofite) de condominios habitacionales en altura en Chile. Cuadernos de Vivienda y Urbanismo, **11**(21) (2018)
12. Comisión Nacional de Energía [CNE]: Consumo energético residencial, público y comercial. [Balance energético] (2010)
13. Ministerio del Medio Ambiente Informe del estado del medio ambiente. Santiago (2011)
14. Romero, J.A.T., Del Carmen, M., Callau, V., Gracia, A.M., Mágicos, L.: Arquitectura Bioclimática Y Urbanismo Sostenible (Volumen I) Ed. Prensas Universitarias De Zaragoza **I**, 354 (2009)
15. Valencia, N.: Primer Lugar en concurso de diseño de vivienda social sustentable en la Patagonia / Aysén, Chile. Plataforma Arquitectura (2018). Accessed 17 Apr 2022. https://www.plataformaarquitectura.cl/cl/758211/primer-lugar-en-concurso-de-diseno-de-vivienda-social-sustentable-en-la-patagonia-aysen-chile

Simulation of the Use of Air-Source Heat Pumps in Different European Cities

Agata Ołtarzewska[1](✉) and Dorota Anna Krawczyk[2]

[1] Doctoral School, Bialystok University of Technology, 15-351 Bialystok, Poland
agata.oltarzewska@sd.pb.edu.pl
[2] Faculty of Civil Engineering and Environmental Sciences, Bialystok University of Technology, 15-351 Bialystok, Poland

Abstract. Air-to-air heat pumps transferred the outside or exhausted air to the space requiring heating or cooling, providing an economic and ecological way to maintain thermal comfort in the building. The aim of the analysis is to estimate for which part of the year the temperature in the building can be kept between the desired temperatures, considering the single-family building without any HVAC system or with an air-source heat pump (ASHP), located in four different European cities. Simulations were carried out using Trnsys software. The results show that the ASHP is able to provide indoor temperatures in the range of 20–24 °C for most of the year in three of the four cities analysed, while in the wider range of the temperature set points (19–25 °C), it can provide almost 100% of the desired temperatures during the year in Bologna or Sevilla, as well as meet expectations in colder climatic conditions (Riga or Poznan).

Keywords: Air-source heat pump · Renewable energy · HVAC systems

1 Introduction

The use of renewable energy sources (RES) has been in the spotlight of the global community for several years [1, 2], and such technologies heave become an integral part of our daily lives. Due to the growing popularity of RES, manufacturers offer a wide range of eco-friendly energy solutions, such as heat pumps (HP), which provide an alternative ecological source of heating and cooling at home [3]. Air-source heat pumps (ASHP) work on the similar principle as air conditioners, except that the former work in reverse compared to the latter. ASHP extracts heat from the outside or exhausted air and transfers it to the compressor of the pump's indoor unit. This process is entirely based on the principle of heat transfer, which is aided by the property of the pump's refrigerant. Compression results in a rapid increase in the refrigerant temperature, creating the necessary conditions to transfer the generated heat energy to the room. This warm air enters the building through a series of fan coil units or "blowers". The process is repeated cyclically, so that the heat from the outside or exhausted air is deliver to the space requiring heating or cooling, thus providing an economic and ecological way of maintaining the thermal comfort in the property [3, 4]. With the growing popularity of air-source heat

X. Chen (Ed.): ICESE 2022, ESE, pp. 164–170, 2023.

pumps, they are also becoming an increasingly common research direction. Although heat pumps are efficient appliances of the future and offer numerous advantages (both from an economic and ecological point of view), they are constantly being optimized and improved, to enable their operation in various building types and in all climatic conditions.

Carroll et al. [5] provided a systematic literature review of air-source heat pumps field studies. They focused on three main aspects: 1) defrost management, 2) ASHP system management and 3) ASHP as a demand-responsive component of a smart grid. They indicated the need for further research to see how smart grid heating solutions contribute (among others) to the decarbonization of energy systems.

Shirani et al. [6] considered the energy efficiency of exhaust air heat pumps under different operating conditions. The simulations showed that lowering the set-point temperature by 1 K reduced the heat demand by about 15% on average and that the SCOP of the HP system increases as the set-point temperature of the building increases. Serey et al. [7] explained the working principle of an air-to-air heat pump, considering the different configurations, and evaluated their performance. They showed that ASHP can be further optimized by using modern temperature, CO_2, wind speed sensors or dumpers and making them "smart systems". Chesser et al. [8] analysed the performance of air-source heat pumps in a deeply retrofitted Irish houses. They proved that all houses qualify as producing renewable heat, but ASHP system results differ from the manufacturer's laboratory test performance. At the higher outdoor temperature, the level of error was: -16% for the 8.5 kW ASHP and -24% in the case of the 11.2 kW device, while at lower temperatures: -3% for 8.5 kW ASHP and -11 for 11.2 kW heat pump. Air-to-air heat pumps were analysed in [9–12], as examples of heat sources in different types of buildings or in various climate conditions and also compared to air-to-water heat pumps in [13], where experimental research and dynamics simulation proved that air-to-air heat pumps have better performance and thermal comfort than air-to-water heat pumps.

As current research focuses mainly on the technical and technological aspects of air-to-air heat pumps or comparing them with other heat sources (e.g. air-to-water or ground source heat pumps), this study fills a gap in verifying the desirability of these devices in four European countries with very different climatic conditions – from the southwesternmost Seville, via Bologna and Poznan, to the northeasternmost Riga. This research aims to estimate for which part of the year the temperature in the building can be kept between the desired temperatures, considering the building with and without an ASHP, located in four different European cities.

2 Materials and Methods

The simulation was carried out for a one-storey residential building, built in traditional technology and characterized by a simple construction. The object consists of a living room with a kitchen, bathroom, bedroom, room and hall. The area of the building was set as 100 m^2, while the cubature – as 270 m^3. The building is intended for the living of 3–4 people (single-family). The materials of which it was made meet all applicable technical and quality standards. Due to higher accuracy and speed of performed calculations compared to software such as EnergyPlus or Audytor OZC, as well as wider availability of weather data from meteorological stations in different parts of Europe and the world, energy simulations of the building model were performed in the TRNSYS Simulation Studio. For this purpose, Type 88 (Single Zone, Lumped Capacitance Building) was used. The building plan is presented in Fig. 1, while the model of the system with an ASHP performed in Trnsys – in Fig. 2.

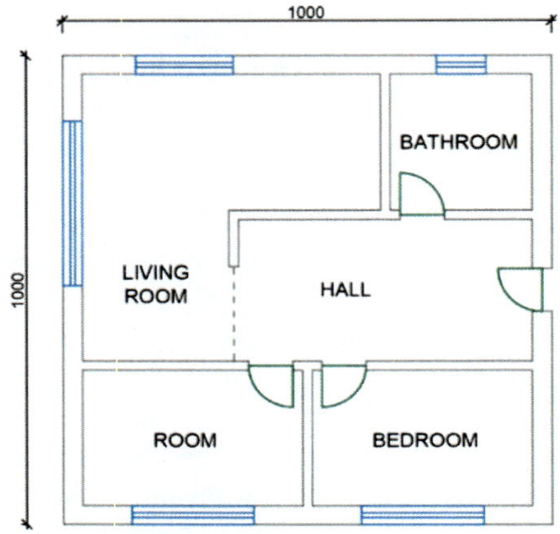

Fig. 1. Plan of the analysed building.

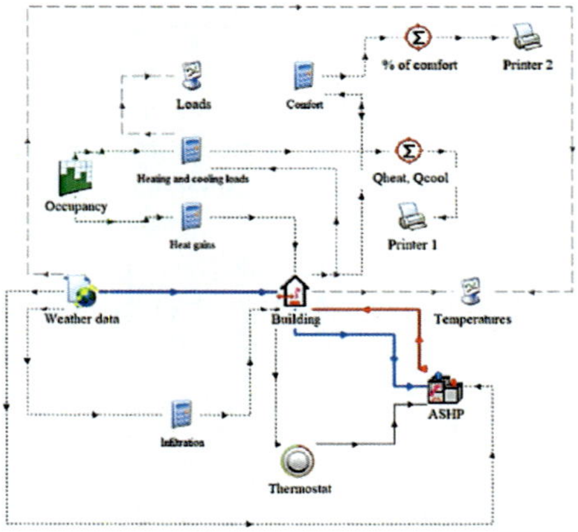

Fig. 2. Model of the system with an air-source heat pump performed in Trnsys.

The heat transfer coefficients "U" of the building envelope were assumed as follows:

- Floor on the ground: U = 0.3 W/m^{2}K,
- Roof: U = 0.15 W/m^{2}K,
- External wall: external wall: U = 0.2 W/m^{2}K,
- Windows and doors: windows and doors: U = 0.85 W/m^{2}K.

Simulations were performed for 4 locations in Europe: Bologna (Italy), Sevilla (Spain), Riga (Latvia) and Poznan (Poland). In order to estimate the energy gains from people, equipment, and lights, an occupancy schedule was set. The indoor temperature in the heating season was set as 20 °C, while in the rest of the year –24 °C. These temperature levels are set by the simple thermostat and should be maintained by a heat pump. The analysed heat source is an air-to-air heat pump, with a total air flow rate of 300 l/s and no auxiliary heating.

According to [14] it has been assumed that thermal comfort is defined as that condition of mind which expresses satisfaction with the thermal environment, i.e. the condition when someone is not feeling either too hot or too cold. The feeling of thermal comfort depends on several factors, such as metabolic rate, clothing insulation, air temperature and speed, or even relative humidity or skin wetness. However, based on the ASHARE Standard 55–2017 [15], in residential buildings, air temperatures can range from about 67 to 82°F (about 19.5 to 27.5 °C) to provide thermal comfort conditions.

The purpose of the analysis is to estimate for which part of the year (8760 h), the temperature in the building is maintained between the desired temperatures, considering the building without any HVAC system and the building with an air-source heat pump and to verify if the air-source heat pump is able to maintain thermal comfort conditions in the analyzed locations.

3 Results and Discussion

The results of the building energy simulations (heating and cooling loads) performed in Trnsys for four analyzed European cities are presented in Fig. 3.

The highest maximum heat load was obtained in Riga (about 15000 kJ/h), while the highest maximum cool load in Sevilla (about 8000 kJ/h). This is due to the fact that Riga, as the northeasternmost of all analyzed cities, has a climate with continental characteristics, where winters are long, cold, and dark, and on the coldest days the temperature can drop as low as -30 °C. The summer season there is short and relatively mild. Seville, on the other hand, is located in the southern part of Europe, where summers are hot, dry, and almost cloudless, and winters are mild.

It can be noticed that between Riga and Poznan, characterized by a similar climatic conditions, the heating, and cooling loads profiles are also comparable. In the case of Bologna, we can observe a significant demand for both heating in the winter season and cooling in the summer. On the other hand, in Sevilla, there is a much higher demand for cooling in the summer than for heating in winter.

The results of the estimation for which part of the year the temperature in the building was maintained between the set temperatures, considering the building with any HVAC system and building with an air-source heat pump were summarized in the Table 1.

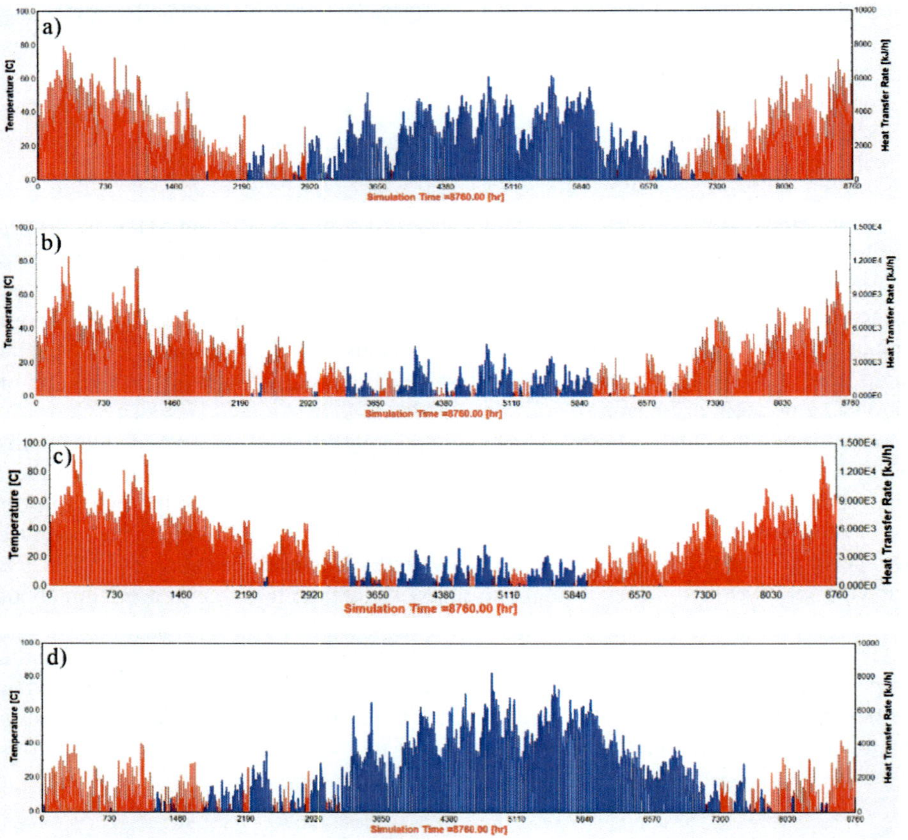

Fig. 3. Results of the building energy simulations (heating and cooling loads) for 4 locations: a) Bologna, b) Sevilla, c) Riga and d) Poznan

Table 1. Results of the estimation for which part of the year the temperature in the building was maintained between 20–24 °C or 19–25 °C

City	without HVAC (20–24 °C)	with ASHP (20–24 °C)	without HVAC (19–25 °C)	with ASHP (19–25 °C)
Bologna	12.9%	57.1%	19.9%	99.4%
Sevilla	21.7%	61.3%	31.1%	99.5%
Riga	14.3%	48.8%	20.6%	85.0%
Poznan	16.2%	54.7%	23.6%	93.5%

Taking into account the temperature level between 20–24 °C, it can be noticed that in the case of Sevilla (Spain), only well-isolated building envelopes provided the required temperature for 21.7% of the year, while in the case of Bologna – only for 12.9%. When the air-source heat pump was used in the building, this value in Sevilla rises to 61.3%, while in Bologna to 57.1% and this is where the largest increase in the analysed value was observed after the implementation of the air-source heat pump. Moreover, if the desired temperature was set between 19–25 °C (only 1 °C difference between the upper and lower set points), air-to-air heat pump was able to provide almost 100% of these temperatures during the year in Bologna and Sevilla, as well as 93.5% and 85.0% in Poznan and Riga, respectively. Extension of the temperature range had the most beneficial effect in the case of Bologna and allowed to increase the analyzed value by 42.3%.

Therefore, the air-source heat pump for heating and cooling proved to be the most advantageous solution in the case of Sevilla, which is characterized by higher ambient temperatures throughout the year, especially during the heating season. Because in the case of Bologna, both systematic heating in summer and cooling in winter are required, less favourable results were achieved there than in Seville.

4 Conclusions

Analysing the results of the simulations, it can be observed that air-to-air heat pump is able to provide an indoor temperature between 20 °C and 24 °C, for most of the year in three of the four cities analyzed. Only considering Riga, the analyzed value did not reach 50%.

In addition, taking into account the wider range of the temperature set points (19–25 °C), air-source heat pump may be able to provide almost 100% of desired temperatures during the year in Bologna and Sevilla (99.4% and 99.5% respectively), as well as meet expectations also in colder climatic conditions (as in Riga or Poznan), maintaining temperatures between 19 °C and 25 °C on approximately 90% of the year.

Therefore, in the light of the presented results, it can be concluded that the air-source heat pump is able to provide relative thermal comfort to "house users" for a major part of the year, regardless of the analyzed location and the desirability of its use in countries with colder climates depends largely on the adopted temperature range (20–24 °C or 19–25 °C), which may reflect individual user preferences.

Acknowledgments. This study was carried out as a part of the work no. WI/WB-IIŚ/9/2022 at the Bialystok University of Technology and was financed from the research subvention provided by the Minister responsible for science.

References

1. Tutak, M., Brodny, J.: Renewable energy consumption in economic sectors in the EU-27. The impact on economics, environment and conventional energy sources. A 20-year perspective. J. Clean. Product. **345**, 131076 (2022)
2. Bórawski, P., Bełdycka-Bórawska, A., Szymańska, E.J., Jankowski, K.J., Dubis, B., Dunn, J.W.: Development of renewable energy sources market and biofuels in the European Union. J. Clean. Prod. **228**, 467–484 (2019)
3. The Pros and Cons of an Air to Air Heat Pump. https://www.greenmatch.co.uk/. Accessed 15 Mar 2022
4. Air-to-air heat pumps. https://energysavingtrust.org.uk/. Accessed 15 Mar 2022
5. Carroll, P., Chesser, M., Lyons, P.: Air source heat pumps field studies: a systematic literature review. Renew. Sustain. Energy Rev. **134**, 110275 (2020)
6. Shirani, A., Merzkirch, A., Roesler, J., Leyer, S., Scholzen, F., Maas, S.: Experimental and analytical evaluation of exhaust air heat pumps in ventilation-based heating systems. J. Build. Eng. **44**, 102638 (2021)
7. Serey, N., Ahmad, D., Jouhara, H.: Air-to-air heat pump: review of recent advances and future potential. E3S Web Conf. **116**, 00074 (2019)
8. Chesser, M., Lyons, P., Padraic, O., Carroll, P.: Air source heat pump in-situ performance. Energy Build. **251**, 111365 (2021)
9. Ala, G., Orioli, A., Di Gangi, A.: Energy and economic analysis of air-to-air heat pumps as an alternative to domestic gas boiler heating systems in the South of Italy. Energy **173**, 59–74 (2019)
10. Chua, K.J., Chou, S.K., Yang, W.M.: Advances in heat pump systems: a review. Appl. Energy **87**(12), 3611–3624 (2010)
11. Tran, C.T., Noël, D., Rivière, P., Arzano, C., Marchio, D.: In-situ method for air-to-air heat pump seasonal performance de-termination including steady-state and dynamic operations. Int. J. Refrig **127**, 239–249 (2021)
12. de Garayo, S.D., Martinez, A., Astrain, D.: Optimal combination of an air-to-air thermoelectric heat pump with a heat recovery system to HVAC a passive house dwelling. Appl. Energy **309**, 118443 (2022)
13. Xiao, B., He, L., Zhang, S., Kong, T., Hu, B., Wang, R.Z.: Comparison and analysis on air-to-air and air-to-water heat pump heating systems. Renew. Energy **146**, 1888–1896 (2020)
14. Thermal comfort in buildings. https://www.designingbuildings.co.uk/. Accessed 15 Apr 2022
15. ANSI/ASHARE Standard 55–2017: Thermal environmental conditions for human occupancy. https://www.ashare.org/. Accessed 15 Apr 2022

Simulation Analysis of Dynamic Characteristics of Urban Central Cooling System

Dong-liang Li[1], Li Gong[2](✉), Guo-qing Zhu[1], Guo-qiang Zhan[3], and Xu e-jie Wang[4]

[1] Power Eng. College, Naval Univ. of Engineering, Wuhan, China
[2] Training Center of Simulation, Naval Univ. of Engineering, Wuhan, China
16898868@qq.com
[3] Teaching and Research Support Center, Naval Univ. of Engineering, Wuhan, China
[4] Zhanjiang PLA91412, China

Abstract. Users' cooling needs also change as the outdoor climate continues to change. In order to ensure the cooling demand and timely grasp the cooling capacity transmission situation and the dynamic characteristics of the heating system of the cooling pipeline network, a simulation model was built to analyze the time delay of the cooling capacity transmission and the influence of the chilled water supply flow rate and temperature on the cooling system. The results show that the chilled water supply temperature has almost nothing to do with the time delay of cold transport, and the chilled water supply flow is the main factor affecting the user's time delay, and the lower the system load, the smaller the chilled water supply flow and the longer the cold transport time. The system load decreases by 10% step, and the cooling capacity of the pipe network, cooling capacity of a chiller, power consumption of the chiller and return water temperature of the cold station gradually decrease with time. The step of chilled water supply temperature increases by 10%, and the step of chilled water supply flow decreases by 10%. The cooling capacity of the system, the cooling capacity of the cooler, the power consumption of the cooler and the return water temperature of the cold station gradually increase with time. Moreover, the system transition time is longer when the chilled water temperature and water supply flow are disturbed simultaneously.

Keywords: Centralized cooling · Thermal system · Simulation model · Cooling capacity

1 Introduction

Rising global temperatures are increasing the load on cooling units and energy losses, thus affecting the cooling effect of the units [1–3]. As there is a lag in the response of the customer's cold load, it is necessary to understand the dynamic change process of the centralized cooling system, and by analyzing the dynamic characteristics, grasp the impact of changes in time, water supply temperature and water supply flow rate on the cooling system, and adjust the cooling capacity of the refrigeration unit in time to better realize the energy saving of the cooling system and reduce the loss of energy consumption.

X. Chen (Ed.): ICESE 2022, ESE, pp. 171–180, 2023.

Domestic and international scholars have investigated the dynamic characteristics of cooling/heating systems. Basak Falay et al. simulate the dynamics of large-scale cooling systems by aggregation methods and reduce the topological complexity of the original network by reducing the number of pipe connections (branches) and pipes, thus reducing the model complexity of centralized cooling systems [4]. Rabah Ismaen et al. proposed a framework for District cooling systems (DCS) analysis by considering inputs, processes (core), and outputs, as well as the structure and behavior of DCS systems to enhance system design and operation and to optimize the overall cost of DCS by integrating system and stakeholder requirements [5]. Jinfu Zheng et al. proposed a functional approach to describe two critical parameters, delay time and relative decay, to represent time delay and heat loss, allowing the pipe network's transient temperature distribution to be obtained [6]. Irina Gabrielaitiene et al. constructed a pipe section model based on the nodal method and applied it to the transient temperature distribution of hot water during transmission and distribution and the short thermal analysis of central heating systems, and then compared the model calculation results with the measured data of the heat source return temperature and the customer water supply temperature to verify the correctness of the model [7]. Prodyut K. Roy and Liu Chao-ying et al. studied the multivariable problem affecting the dynamic change of pipe network with the method of fuzzy control [8, 9].

In the research process, many scholars have conducted a lot of research on design optimization [10–13], cooling load prediction [14–17], and system energy efficiency of the cooling system [18–21], and few have analyzed the dynamic characteristics of the cooling system. Therefore, the article takes a city cooling system as the research object and analyses the impact of the time lag in the process of cold delivery and the perturbation of chilled water supply temperature and supply flow rate on the operation of the city cooling system by establishing a simulation model.

2 System Model

2.1 Pipe Network Simulation Model

In the centralized cooling system, the cooling pipe network plays the role of transporting cold [22]. The network contains two shopping malls and three office buildings. The solid line indicates the water supply pipeline, and the dotted line indicates the return pipeline, the distribution of the cooling pipe network is shown in Fig. 1, and the main parameters of the pipe network are shown in Table 1.

Calculation of the cold capacity delivered by the pipe network:

$$Qgw = \rho Cw(th1 - tg1)Q1{-}2 \tag{1}$$

In the formula: ρ for the density of water, take 1000kg/m^3; $tg1$ for node 1 water supply temperature, °C; $th1$ for node 1 return water temperature, °C; $Q1{-}2$ for the flow rate of pipe Sect. 1–2, m^3/s; Cw for the specific heat capacity of water, take 4.2 kJ/kg·°C.

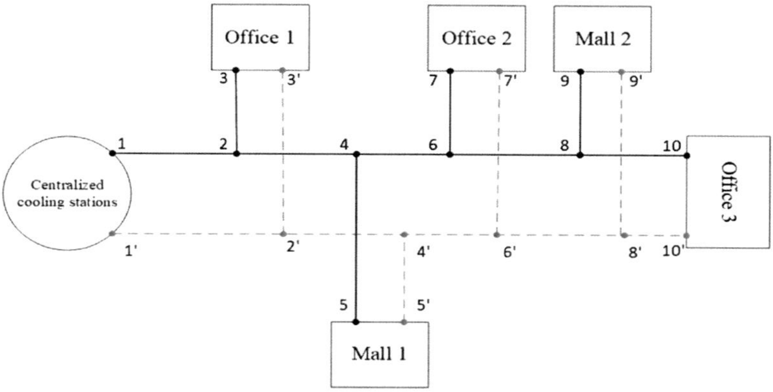

Fig. 1. Schematic diagram of cooling network distribution.

Table 1. Table type styles

Pipe section number	*Pipe length(m)*	*Pipe diameter(mm)*
1–2	500	400
2–4	400	350
4–6	300	300
6–8	400	300
8–10	300	250
2–3	400	200
4–5	350	200
6–7	400	200
8–9	300	200

Return water temperature calculation:

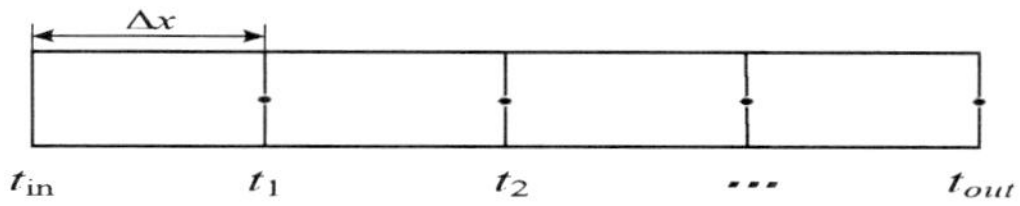

Fig. 2. Distribution diagram of n discrete pipe segments

For the calculation of backwater temperature, a pipe segment of any length L is discretized into an n-segment micro-element pipe segment of length Δx by referring to the pipe segment axial temperature calculation method, as shown in Fig. 2. The model for calculating the return water temperature for each discrete pipe section is shown below.

$$C_{zw}\frac{dt_1}{d\tau} = c_{p,w}G(t_{in} - t_1) - \frac{\pi \cdot \Delta x}{R_{wp}}(t_1 - t_{p,1})$$
$$C_{zw}\frac{dt_2}{d\tau} = c_{p,w}G(t_1 - t_2) - \frac{\pi \cdot \Delta x}{R_{wp}}(t_2 - t_{p,2}) \tag{2}$$

$$C_{zw}\frac{dt_n}{d\tau} = c_{p,w}G(t_{n-1} - t_n) - \frac{\pi \cdot \Delta x}{R_{wp}}(t_n - t_{p,n})$$

In the formula: C_{zw} is the total heat capacity of water, kJ/kg·°C; t_{in} is the inlet fluid temperature of the pipe section, °C; C_{pw} is the specific heat capacity of chilled water, kJ/kg·°C; R_{wp} is the heat transfer thermal resistance of the pipe, (m·°C)/W; G is the water supply flow rate, m^3 /s.

2.2 Chiller Model

The primary function of the chiller is to generate cold capacity. With electricity as the power source, the chiller provides cold capacity for the chilled water to the end of the air conditioner through the refrigerant's heat absorption/heat release effect, releases heat to the cooling water, and dissipates heat from the cooling tower [23].

Chiller plant cooling capacity calculation:

$$\dot{Q}met = COP \cdot Pchiller \tag{3}$$

In the formula: COP is the chiller energy efficiency.

Calculation of actual operating power consumption of chillers:

$$\begin{aligned} Pchiller &= a + b(tcw, out - tchw, in) \\ &+c(tcw, out - tchw, in)^2 + dQ \\ &+eQ^2 + f(tcw, out - tchw, in)Q \end{aligned} \tag{4}$$

In the formula: tcw, out is the cooling water return temperature, °C; $tchw, in$ is the chilled water supply temperature, °C; Q is the ideal cooling capacity of the chiller, kw; *abcdef* is the chiller operating data fitting parameter.

2.3 Time Lag Analysis Model

When the cooling station supplies the cooling to the user, the cooling load changes at any time. In order to meet the needs of the user, the output power of the cooler should be adjusted in time. However, the cooling distance is long, leading to a time delay in the pipeline network in conveying the cooling volume [24].

Pure lag time for cold users n **(**$n = 1, 2, 3, 4, 5$**)** τn **Calculation:**

$$\tau \mathrm{n} = \frac{lin, n}{vin, n} + \sum_{i=1}^{n} \frac{li}{vi} \tag{5}$$

In the formula: lin, n the length of the water supply branch pipe between the cold user n inlet and the primary side of the heat exchanger, m; vin, n is the chilled water flow rate of the water supply branch pipe between the cold user n inlet and the primary side of the heat exchanger, m/s; li is the length of the pipe section between the cold user $i - 1$ and the cold user i on the water supply pipe, m; vi is the chilled water flow rate of the pipe between the cold user $i - 1$ and the cold user i on the water supply pipe, m/s.

3 Simulation Results and Analysis

3.1 Influence of Chilled Water Supply Temperature on Time Lag of Customer Cooling Capacity

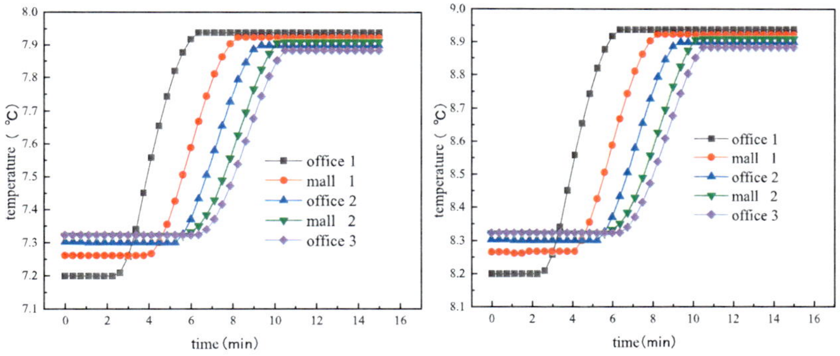

(a) The water supply temperature is increased by 1℃

(b) The water supply temperature is increased by 2 ℃

Fig. 3. Chilled water supply temperature increases by 1 °C, 2 °C cooling capacity lag time of each user

Figure 3 show the lag time for each user's inlet water supply temperature to reach stability when the chilled water supply temperature rises by 1 °C and 2 °C, respectively, under typical working conditions of the pipe network.

From the analysis of Fig. 3a, it can be concluded that the No. 1 office building is the first to reach stability, with a transition time of about 6 min, while the No. 2 shopping mall and No. 3 office building are the last to reach stability, with a transition time of about 10 min. This is because office building No. 1 is the closest to the cold source, while shopping mall No. 2 and office building No. 3 belong to the most terminal users due to the extended supply distance, resulting in a longer chilled water delivery time in the pipe, and thus the transition time when reaching stability is longer. Figure 3b is almost identical to the dynamic change curve in Fig. 3a. Therefore, the chilled water supply flow rate remains unchanged and changing only the chilled water supply temperature has little effect on the time lag of the user's cooling capacity.

3.2 Effect of Chilled Water Supply Flow on the Time Lag of Customer Cooling Capacity

The chilled water supply flow rate of 859m^3/h was selected as the 100% flow rate condition, and the lag time for each user's inlet water supply temperature to reach stability was analyzed under 100% and 75% flow rate conditions, with the chilled water supply temperature rising by 1 °C, as shown in Fig. 4.

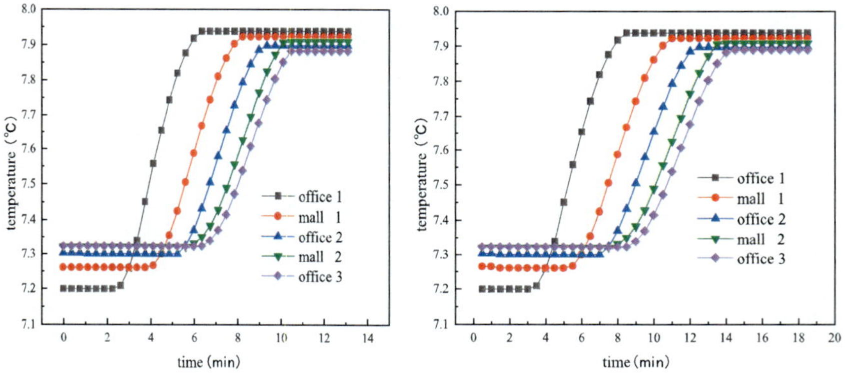

(a)100% water supply flow rate (b)75% water supply flow rate

Fig. 4. Cooling capacity lag time of each user under different flow conditions

From Fig. 4, it can be visually analyzed that under the 100% flow condition, office building No. 1 is given priority to reach stability with a transition time of about 6 min, office building No. 2 as an intermediate user with a transition time of 9 min, shopping mall No. 2 and office building No. 3 are the last to reach stability with a transition time of about 10 min. In comparison, the lag time under the 75% flow condition is longer. Because the chilled water flow condition is minor, the chilled water supply flow rate in the pipe is small because the supply cooling distance remains the same, the cold quantity from the refrigeration station to the user of a longer time, and the shorter the flow rate, the longer the lag time.

Therefore, the chilled water supply flow rate mainly affects the time lag of the customer's cold quantity. The smaller the supply flow rate, the longer the mean quantity delivery time, and the greater the impact on the time lag of the customer.

3.3 Dynamic Analysis of Changing the Chilled Water Supply Temperature

Table 2 shows each parameter's magnitude of change and transition time before and after the disturbance when the chilled water supply temperature rises by 10%. From Table 2, it can be analyzed that when the chilled water supply temperature rises by 10%, the cold volume delivered by the pipe network, the chiller cooling capacity, the chiller power consumption and the cold station return water temperature have a sudden increase in the process within 20min, the increase is 8.2%, 10.2%, 8.6%, 3.6% respectively because when the chilled water supply temperature rises suddenly, the user end water supply temperature will rise slowly in a period, resulting in the supply and return water The temperature difference increases, thus the heat exchange between the evaporator of the chiller and the chilled water increases, the cooling capacity of the chiller increases and the power consumption of the chiller increases.

Table 2. Table of main parameters of water supply temperature disturbance

Projects	Unit	Pre-disturbance	Post-disturbance	Change/%	Transition time/min
Chiller cooling capacity	kW	4516.3	4887.3	8.2	20
Pipeline ne work delivery of cold capacity	kW	4240.6	4671.2	10.2	
Chiller power consumption	kW	817.5	888.1	8.6	
Return water temperature	°C	12.045	12.474	3.6	

3.4 Dynamic Analysis of Changing Chilled Water Supply Flows

Tab.3 shows each parameter's magnitude of change and transition time before and after the disturbance when the chilled water supply flow rate is reduced by 10%. From Tab.3, it can be found that when the water supply flow rate is reduced by 10%, the system delivery cold volume, chiller cooling capacity, unit power consumption and cold station return water temperature have a sudden rise in the process within 10min, with increases of 8.6%, 10.7%, 10% and 4.7% respectively. When the chilled water supply flow drops suddenly, the end-user return water temperature rises rapidly within a short period, increasing the temperature difference between the chiller supply and return water, thus increasing the heat exchange between the chiller evaporator and the chilled water, the chiller cooling capacity increases, the pipe network conveys more cooling capacity and the chiller power consumption increases.

Table 3. Table of changes in the main parameters of water supply flow disturbance

Projects	Unit	Pre-disturbance	Post-disturbance	Change/%	Transition time/min
Chiller cooling capacity	kW	4690.1	5095.7	8.6	10
Pipeline network delivery of cold capacity	kW	4464.4	4970.2	10.7	
Chiller power consumption	kW	849.4	934.5	10.0	
Return water temperature	°C	12.045	12.607	4.7	

3.5 Dynamic Analysis of Simultaneous Changes in Chilled Water Supply Temperature and Supply Flow Rate

When the chilled water supply temperature increases and the chilled water flow rate decreases, there is an abrupt increase in the unit cooling capacity, pipe network delivered cooling capacity, unit power consumption, and chilled water return temperature within 16 min, with increases of 18.9%, 22.3%, 18.6%, and 7.9% respectively. The analysis shows that, compared to the previous perturbation analysis of individual factors, the simultaneous perturbation of two factors approximately doubles the increase.

4 Conclusions

The main factor affecting the user's time lag is the chilled water supply flow. Chilled water supply temperature is almost irrelevant to the time lag of cold transport, and the lower the system load, the smaller the chilled water supply flow and the longer the mean transport time.

The chilled water supply temperature rises by 10%, the water supply flow rate decreases by 10%, and the system transported cold quantity, chiller cooling capacity, chiller power consumption, and cold station return water temperature gradually increases with time. And chilled water supply temperature and water supply flow rate at the same time perturbation, the system change amplitude is more excellent.

By constructing a simulation model of the cooling system and mastering the dynamic change process of the cooling system, it can not only adjust the amount of cooling in time, better guarantee the cooling supply, but also avoid the loss of cooling capacity and realize the energy saving of the cooling system, which has certain practical guiding significance. However, in the research process, the simulation model established is only suitable for the dendritic pipe network of a single cold source or multiple cold sources and is not suitable for the ring cooling pipe network, which has certain limitations.

References

1. Østergaard, P. A., et al.: The four generations of district cooling - A categorization of the development in district cooling from origin to future prospect Energy, 253, p. 124098 (2022)
2. Zhang, Y., Liu, Y., Liu, Y., Jiang, B.: A Hybrid Dynamical Modelling and Control Approach for Energy Saving of Central Air Conditioning Math. Probl. Eng. vol. 2018, p. 6389438 (2018)
3. Long, W., Bai, W., Fan, R., Liang , H., Liu, K.: After the epidemic: Energy bus and the development of the 5th generation district heating and cooling system," HVAC, vol.50 (10), pp.1–13, (2020)
4. Falay, B., Schweiger, G., Donovan, K.O., Leusbrock, I.: Enabling large-scale dynamic simulations and reducing model complexity of district heating and cooling systems by aggregation. Energy **209**, 118410 (2020)
5. Ismaen, R., El Mekkawy, T.Y., Pokharel, S., Al-Salem, M.: System requirements and optimization of multi-chillers district cooling plants. Energy **246**, 123349 (2022)
6. Zheng, J., Zhou, Z., Zhao, J., Wang, J.: Function method for dynamic temperature simulation of district heating network. Appl. Therm. Eng. **123**, 682–688 (2017)
7. Gabrielaitiene, I., Bøhm, B., Sunden, B.: Modelling temperature dynamics of a district heating system in Naestved, Denmark—A case study. Energy Convers. Manage. **48**, 78–86 (2007)
8. Y. L. C., L. S. X., Y. S. Z., C. S. Y.: Control System Design of Heat Exchange Station Based on Fuzzy Technology, In: 2006 International Conference on Machine Learning and Cybernetics, pp. 380–384 (2006)
9. K. R. P., K. I. M. G., C. H. B.: Fuzzy rule-adaptive model predictive control for a multivariable heating system," In: Proceedings of 2005 IEEE Conference on Control Applications, 200CCA, pp. 260–265 (2005)
10. Neri, M., Guelpa E., Verda, V.: Design and connection optimization of a district cooling network: Mixed integer programming and heuristic approach Appl. Energy, vol. 306, p. 117994, (2022)
11. Schweiger, G., Larsson, P., Magnusson, F., Lauenburg, Velut, S.: District heating and cooling systems – Framework for Modelica-based simulation and dynamic optimization, Energy, 137, pp. 566–578, (2017)
12. Abugabbara, M., Javed, S., Johansson, D.: A simulation model for the design and analysis of district systems with simultaneous heating and cooling demands, Energy, vol. 261, p. 125245, (2022)
13. Hinkelman, k. et al.: Modelica-based modeling and simulation of district cooling systems: A case study," App. Energy, 311, p. 118654, (2022)
14. Wang, L., Lee, E.W.M., Yuen, R. K. K.: Novel dynamic forecasting model for building cooling loads combining an artificial neural network and an ensemble approach, Appl. Energy, 228, pp. 1740–1753, (2018)
15. Rana, M., Sethuvenkatraman, S., Goldsworthy, M.: A data-driven approach based on quantile regression forest to forecast cooling load for commercial buildings Sustain. Cities Soc. 76, p. 103511, (2022)
16. Gang, W., Wang, S., Shan, K., Gao, D.: Impacts of cooling load calculation uncertainties on the design optimization of building cooling systems, Energy Build. 94, pp. 1–9, (2015)
17. Zhang, F. , Hou, Z., Luo, Y.: Optimal operation strategy of ice storage air conditioning system based on load forecasting, Refrigeration and Air Conditioning, 20, pp.65–70, (2020)
18. P. K. Sahoo, "Exergoeconomic analysis and optimization of a cogeneration system using evolutionary programming," Applied Thermal Engineering, vol. 28, pp. 1580–1588, 2008–01–01 2008

19. Guo, C., Chu, S., You, Y.: Theoretical and experimental study on Energy efficiency of Regional Cooling System," J. Tianjin Chengjian University, 24, pp.62–67, (2018)
20. He, A., Xi, H., Ding, T., Wang, J., Li, Z.: Energy efficiency optimization of an integrated heat pipe cooling system in data center based on genetic algorithm, App. Thermal Eng. vol. 182, p. 115800, (2021)
21. Lyu, W., et al.: Energy efficiency and economic analysis of utilizing magnetic bearing chillers for the cooling of data centers," J. Build. Eng., 48, p. 103920, (2022)
22. Fan, Z., Xiong, D., Yang, L., Rong, X., Chen, J.: Optimal Design of Pipe Diameter of regional cooling and Heating system," HVAC, 51(08), pp.50–55, (2021)
23. Li, Z., Guo, H.: Multi-model Fusion of Central Air conditioning chiller, Comput. Simul. 39, (08), pp.311–316, (2022)
24. Kang, Y., Zuo, Z.: Cooling Loss Analysis of secondary pipe network of regional cooling system," HVAC, 39 (11), pp.31–36, (2009)

Author Index

A
Alariqi, Mona 144
Almendrala, Michelle 29

C
Carmona, Samuel 158
Chindamanee, Phacharawat 151

D
Deng, Panpan 131
Ding, Xiaobin 118
Doma, Bonifacio 29

E
Espinoza, Paola Jeremias 3
Evidente, Ralph Carlo 29

F
Flores, David 158
Fu, HaiYan 17
Fu, Haiyan 50, 87
Fujii, Manabu 59, 103

G
Gong, Li 171
Guo, Jian 118

H
Hinostroza, Steve Dann Camargo 94

I
Ibrahim, Mona G. 59, 103

K
Krawczyk, Dorota Anna 164

L
Li, Chunying 37
Li, Dong-liang 171
Li, Shuyong 118
Li, Zhenyu 74
Liu, Binghui 87
Long, Wei 144
Luo, Jiajun 17

M
Mansour, Mahmoud M. 103
Martinez, Nelida Tantavilca 3
Musah, Baba Imoro 144

N
Nasr, Mahmoud 59, 103

O
Ołtarzewska, Agata 164

P
Pan, Yingdong 118
Pareti, Stefania 158
Pichetjamroen, Achara 151
Ponce, Magaly Godys Avellaneda 94

R
Rudolph, Loreto 158

S
Samaniego, Anali Chulluncuy 94
Saqr, Ahmed M. 59
Somakettarin, Natthawuth 151

T
Tantavilca, Claudio Lopez 3
Teerakawanich, Nithiphat 151
Tizon, Zhane Ann 29

X. Chen (Ed.): ICESE 2022, ESE, pp. 181–182, 2023.

V
Villaflor, Shekinah Mae 29

W
Wang, ShuGuang 17
Wang, Xu e-jie 171
Wu, Feili 118
Wu, YiCheng 17
Wu, Yicheng 50, 87

X
Xing, Lifeng 74

Y
Yao, Huichao 50
Yoshimura, Chihiro 59, 103
Yu, Chen 50, 87

Z
Zevallos, Deysi Aliaga 3
Zhan, Guo-qiang 171
Zhang, Chunlong 37
Zhang, Lukai 74
Zhang, Peng 74
Zheng, YuanMao 17
Zhou, Zhouyi 87
Zhu, Guo-qing 171